T0205593

Lecture Notes in Networks and Systems 677

The series "Lecture Notes in Networks and Systems" publishes the latest developments in Networks and Systems—quickly, informally and with high quality. Original research reported in proceedings and post-proceedings represents the core of LNNS.

Volumes published in LNNS embrace all aspects and subfields of, as well as new challenges in, Networks and Systems.

The series contains proceedings and edited volumes in systems and networks, spanning the areas of Cyber-Physical Systems, Autonomous Systems, Sensor Networks, Control Systems, Energy Systems, Automotive Systems, Biological Systems, Vehicular Networking and Connected Vehicles, Aerospace Systems, Automation, Manufacturing, Smart Grids, Nonlinear Systems, Power Systems, Robotics, Social Systems, Economic Systems and other. Of particular value to both the contributors and the readership are the short publication timeframe and the world-wide distribution and exposure which enable both a wide and rapid dissemination of research output.

The series covers the theory, applications, and perspectives on the state of the art and future developments relevant to systems and networks, decision making, control, complex processes and related areas, as embedded in the fields of interdisciplinary and applied sciences, engineering, computer science, physics, economics, social, and life sciences, as well as the paradigms and methodologies behind them.

Indexed by SCOPUS, INSPEC, WTI Frankfurt eG, zbMATH, SCImago.

All books published in the series are submitted for consideration in Web of Science.

For proposals from Asia please contact Aninda Bose (aninda.bose@springer.com).

Tokuro Matsuo · Takayuki Fujimoto ·
Ford Lumban Gaol
Editors

Innovations in Applied Informatics and Media Engineering

Editors
Tokuro Matsuo
Advanced Institute of Industrial Technology
Tokyo, Japan

Takayuki Fujimoto
Toyo University
Kawagoe, Japan

Ford Lumban Gaol
Doctor Program of Computer Science
Bina Nusantara University
Jakarta, Indonesia

ISSN 2367-3370 ISSN 2367-3389 (electronic)
Lecture Notes in Networks and Systems
ISBN 978-3-031-30768-3 ISBN 978-3-031-30769-0 (eBook)
https://doi.org/10.1007/978-3-031-30769-0

This Springer imprint is published by the registered company Springer Nature Switzerland AG
The registered company address is: Gewerbestrasse 11, 6330 Cham, Switzerland

Preface

This book includes theory and practice on Applied Informatics and Media Engineering from an academic and professional viewpoint. Contents of this book are separated into three groups; (1) Applications in Applied Informatics and Media Engineering (Chapters 1–5), (2) Sociological, Psychological and Philosophical Aspects in Applied Informatics and Media Engineering (Chapters 6–11) and (3) Theory and Elemental Technology in Applied Informatics and Media Engineering (Chapters 12–15). Chapter 1 presents electronic book application based on users' experience in real-life. Chapter 2 presents a support system to aid traveler's activity through location information. Chapter 3 presents the drawing software with the drawing feeling for users. Chapter 4 presents a video discussion system for university students for online learning. Chapter 5 designs an application that can simulate the experience of the Theatre Optique to enhance the creativity. Chapter 6 surveys and analyzes the existing chatbots of various scales and their services. Chapter 7 shows the philosophical issues in post-Internet age in creativity point of view. Chapter 8 focuses on the 'page-turning' motion as a factor that enhances the reading experience and conducts experiments and verifications. In Chapter 9, authors focus on the design and color in real city and scene to understand the well-designed landscape. Chapter 10 presents and discusses non-pharmacological treatment for dementia in Applied Informatics point of view. Chapter 11 describes the proper methods to aid the addicted people to video game. Chapter 12 is a contribution that investigates the ontology in regard with the traffic accidents. Chapter 13 proposes an approach to partition large ontology for change effects managing, and it consists of creating a weighted dependency graph from ontology structure and then determining communities using the Island Line algorithm. Chapter 14 is a contribution of outcomes of digital marketing for small-middle size business organizations. Chapter 15 shows a combination of bidirectional associative memory and backpropagation neural network and training process of the signature in the form of digital images.

Dr. Matsuo, Dr. Fujimoto and Dr. Lumban Gaol are grateful to the authors and the reviewers for their contribution to this work. Editors also acknowledge with their gratitude the editorial team of Springer Verlag for their support during the preparation of the manuscript.

February 2023

Tokuro Matsuo
Takayuki Fujimoto
Ford Lumban Gaol

Contents

Analog on Digital (AoD) Theory-Based Design of Book Application for Representing Real-Life Experience 1
Ziran Fan and Takayuki Fujimoto

The Modelling and Designing of Tourism Activity Locator Application 14
Ford Lumban Gaol, Tommy Wijaya, and Tokuro Matsuo

Development of Drawing Software that Gives a Staple Drawing Feeling by the Difference in Sound 22
Yulana Watanabe and Takayuki Fujimoto

The Impact of Video Conference Application to College Students in Online Learning Activities 37
Givbrela Lostawika, Devi Siti Azzahara, Marven Immanuel Christianto, Nunik Afriliana, Tokuro Matsuo, and Ford Lumban Gaol

Proposal for a Theatre Optique Simulated Experience Application 48
Nanami Kuwahara and Takayuki Fujimoto

A Classification and Analysis Focusing on Attempts to Give a Computer a Personality: A Technological History of Chatbots as Simple Artificial Intelligence 59
Taishi Nemoto and Takayuki Fujimoto

With Post-internet Society, *The Third Agent* 71
Takashi Shimizu

Research on Digital Reading Experience with 'PAGE-Turning' Physical Feedback 80
Yulana Watanabe and Takayuki Fujimoto

Method of Extracting Community Colors with Local Characteristics in Landscape Planning: Planning the Color Scheme for the Taisetsu-Furano Route of the Scenic Byway Hokkaido 91
Kasai Daisuke and Miyauchi Hiromi

Non-pharmacological Treatment of Dementia from the Perspective of Applied Informatics 101
Ken-ichi Tabei

Research Methods to Build a Reference Model for Designing Addiction-Aware Video Game 111
Flourensia Sapty Rahayu, Lukito Edi Nugroho, and Ridi Ferdiana

TrAcOn: A Traffic Accident Ontology for Identifying Accidents-Prone Areas in Senegal 121
Mouhamadou Gaye, Ibrahima Diop, Ana Bakhoum, and Papa Alioune Cissé

Ontology Partitioning for Managing Change Effects 132
Mouhamadou Gaye and Ibrahima Diop

The Effect of Digital Marketing on Micro, Small and Medium Enterprise in Indonesia 147
Alvin Igo Sasongko, Gregorius Christian Widjaja, Jason Theodore, Nunik Afriliana, Tokuro Matsuo, and Ford Lumban Gaol

Bidirectional Associative Memory as Normalisator Backpropagation Neural Network in the Signature Image Training 157
Fransisca Fortunata Dewi, Ford Lumban Gaol, and Tokuro Matsuo

Author Index 165

Analog on Digital (AoD) Theory-Based Design of Book Application for Representing Real-Life Experience

Ziran Fan(✉) and Takayuki Fujimoto

Graduate School of Information Sciences and Arts, Toyo University, Tokyo, Japan
swterc@gmail.com, fujimoto@toyo.jp

Abstract. This study discusses digitization in today's society and examines the relation be-tween digital and analog in terms of users' content experience. As a result of our analysis, we focused on the "reality" which analog devices deliver. In this paper, we devise a design method that reflects reality, an analog feature, into digital content based on Analog on Digital (AoD) theory. The study aims to construct a book application that represents real-life experience by designing the physical sensations of reading paper books. The application represents the stains and wrinkles of paper books in the application interface, being inspired by the users' experience with analog devices whose condition changes as they are used. On top of this, we incorporate in the application the sensations users get when tilting a book to open it and devise a method to move through pages by detecting the angle and speed of tilting the device.

Keywords: Reading Experience · Analog-like Reality · Digital Contents · Application Design. · Interaction Design

1 Introduction

1.1 Background of Digitization

Recent advancement in digitization has been remarkable. Functionality of digital devices like smartphone has improved over the years, making our life more convenient. Moreover, application of next-generation technologies such as VR, 5G and IoT has been promoted in society at a steady pace. Today, our lifestyle is supported by digital technologies, and digitization has brought us many benefits.

Smartphone is a typical example of digitization. Most of our daily activities can now be performed using a single smartphone device. According to Appfigures report for 2018, the number of Apple's iOS smartphone apps is 2.2 million, and that of Google's Android apps is 3.6 million [1]. Moreover, hundreds of thousands of apps are newly released every year.

Tens of millions of apps cover all the daily activities such as work, home life, education and hobby. People not only obtain information via smartphone device but also sort and deliver information through it. Smartphone has become a necessity of life that people take with them everywhere.

T. Matsuo et al. (Eds.): AIMD 2019, LNNS 677, pp. 1–13, 2023.
https://doi.org/10.1007/978-3-031-30769-0_1

1.2 Definition of Digital and Analog

The word digital has multiple meanings; it originally refers to presentation of information by continual numerals. The idea of analog, which is the opposite of digital, means an information format continuously represented by simulated physical quantity [2]. However, current digital devices employ both digital and analog forms of information presentation, and the difference between these two is becoming unclear, especially in digital content design. Therefore, this study attempts to define digital and analog in terms of user experience of the content instead of the viewpoint of information presentation.

Digital:

provides data content, e.g., applications and software, based on electronic structure to deliver a virtual user experience

Analog:

functions to provide physical experience, e.g., machines and instruments, based on physical principles

1.3 Demand for Analog

While our lifestyle is being further digitized, the demand for conventional mechanical or physical analog devices and tools is still high. For example, according to a report released by Recording Industry Association of Japan, 1,003,934 units of audio records were sold in 2018 and the sales reached $1.5 billion. On the other hand, the sales of music data distribution service via downloads and streams were $609 hundred million. The analog disk market is especially significant in that it has increased for 5 years in a row [3]. The data shows that people prefer the physical media such as analog records and disks to music data distribution service though the latter excels in convenience.

In addition, in publishing industry, the size of paper book market is still huge. Book applications are now pre-installed on smartphones and tablets, and tens of thousands of books could be stored on one device. Moreover, electronic devices like Kindle by Amazon and Reader by Sony are intended for the use by electronic book readers and are rich in reading functions. However, in the current publishing industry, while the size of the market of both paper books and electronic books reaches $7.3 billion, electronic books account for only 17.7% at $1.3 billion [4]. 2010 is called the first year in the history of electronic books for the release of Kindle by Amazon and iPad, the tablet device by Apple. Although a decade has passed since then and the usage and awareness of electronic books have grown over the years, they are still not in a position to outdo paper books in either demand or market share [5].

1.4 Analog Feature of "REALity"

Digital devices like smartphone excel in convenience and functionality, and are now used so widely that everybody has one. Nevertheless, people tend to prefer using analog devices to digital counterparts. The reason lies in the analog features.

As mentioned above, analog is defined as "something that functions based on physical principles to provide physical user experience, e.g., machines and instruments". Analog devices like books and disks are tangible, and you can actually touch and feel them. The sensation conveyed through your body is an essential factor in perceiving and using objects [6].

Sensory experience with analog objects also exists in the process of using them not just in their presence. Because digital content such as smartphone apps provides virtual experience, users could not feel the touch of a real object or experience a sensation of operating it. For example, using the keys on the physical keyboard to enter text gives users a better sensation of typing interaction than using the touchscreen keyboard on smartphone.

Also, the interface on current smartphones is putting touch gestures at the center of its design, and similar gesture operations like tapping, swiping and scrolling are adopted in any applications [7]. In contrast to this, people use analog devices with many different operations like turning, twisting, turning over, folding and combining. However, it is difficult to realize these analog operations on digital devices such as smartphone. The diverse operations users perform when using an object are replaced by the functions of on-screen interface buttons and are simplified into touch gesture operations.

This is why analog devices are more effective in providing the sensations of using objects than digital devices. This realistic experience corresponds with human physical sensations. Such reality creates joy in the use of an object and improves users' willingness to and motivation at work [8].

Reality, which is inherent in physical objects, explains the never-diminishing demand for analog objects. People find values more in audio records and disks than in the music available as audio data, and would rather read a real book, touching the paper, than on an electronic device. This is a natural tendency that people generally have. "Reality", a feature of analog devices, could never be replaced by the convenience and functionality of digital devices no matter how excellent they are. People would rather attempt to utilize both digital and analog advantages. In fact, digital devices are often used along with analog devices [9].

In other words, it also indicates the need for a design method that merges digital convenience with analog reality. In this paper, we consider the possibility of a design method that incorporates reality as an analog feature into digital content. In an attempt to achieve this, we take paper books and electronic books mentioned above as an example and propose a digital book application that represents the realistic sensations of paper books.

2 Design Theory

2.1 Related Studies

There are many conversion examples of analog to digital. As for smartphones, it is clear that approximately all the applications are designed based on the models of analog devices, tools or contrivances. Especially the interface of analog devices is often reproduced on smartphones, and numerous applications digitized the everyday-life analog tools such as the clock, calendar, notebook and camera.

Apart from the interface, some precedent studies aimed to represent the analog sensations on digital devices. For example, a simulating system proposed in 1998 represented the tactile sensation and the feeling of weight of virtual objects on the force controller put on the finger by calculating the value of friction, hardness, gravitational acceleration and interaction of fingers when touching objects [10].

As another example, in 2018, Nintendo, a game company, released Nintendo Labo, which is an entertainment product incorporating analog operational sensations into game, and attracted public attention. Nintendo Labo provides users with the realistic sensations as if the game users were actually viewing with the telescope, hooking a fish, playing the piano or driving a car in the game by combining the controller, 'Switch', which is a game console, and the cardboard crafts of telescope, fishing rod, piano and steering wheel. The system detects users' movements by the gyro sensor equipped with the controller and reflects them into the game [11].

Thus, the attempts to design content by representing the analog elements on digital devices have been conducted with the multilateral approach. However, there is no design method that clearly claims the reflection of analog reality to the content design and integrated concept to merge the analog features to digital advantages yet as far as the authors examined.

2.2 Design Theory of Analog on Digital (AoD)

Analog on Digital (AoD) is the design theory proposed by Fujimoto (2018). It refers to the concept incorporating analog elements and features into digital content and using digital devices with the analog-like approaches [12]. This idea is opposite to the Internet of Things (IoT), the information technology of coming generation, which is expected to construct the information environment of the society. IoT aims at the extreme digitization that applies digital technologies to all things by sensor network. For the purpose, the applicability for society environment and practical user convenience are considered remarkably crucial in the idea of IoT [13].

In contrast to IoT, AoD focuses on the user experience regarding the use of the technology, in other words, the sensory experience of using tools. As being represented with the instance of IoT, today digital technologies tend to seek for the improvement of functionality. Even though the convenience of people's daily life is enhanced, the sensory experience of touching, feeling and using something is going to be faded. To represent reality on digital devices, recently a lot of content have been provided with Augmented Reality (AR) or Virtual Reality (VR), however, the technical problems obstructing the general use for the everyday-life activities have been left and it makes those content products just the expression like games or arts [14].

The representation of sensory reality considered in AoD focuses on the new usage design of existing digital content and devices instead of the utilization of advanced technologies such as AR and VR. AoD incorporates the analog usage into digital devices as IoT embeds sensors into things to use them like digital devices. In respect of this point of view, it can be said that, the example of Nintendo Labo mentioned above corresponds to the idea of AoD.

The purpose of AoD is to design something that users want to use rather than providing something just for users' convenience, i.e., 'something easy to use.' For that, the reality of using tools is the essential factor [12]. This study applies the AoD design theory to the smartphone application design, and designs the digital book application representing sensory reality.

3 Design of Digital Book Application to Produce Sensory Experience

3.1 Design Overview – Representing "THe Sensory Reality of Reading Book"

According to the survey by Yaguchi, Uemura (2011), regarding the users' comments to electronic books, most users thought that they are not satisfactory in reading experience, and the results indicated that electronic books are inferior to paper books in respect of user experience [5].

Electronic books resemble paper books regarding the interface design and they represent not only the surface design but also the animation expression of turning over pages with the movements of paper. Moreover, changing the text character size, background image and screen brightness is also possible. Electronic books excel in the functions that support reading [15].

However, the electronic books lack the representation of the user's interaction with the book as a tool. For example, the wrinkles would be produced with the long-term use regarding paper books and a user can open books easier with them, on the other hand, that experience is not represented in electronic books. The remarkable digital feature is the constant condition of content despite of the frequent use, however, in the other words, it has no reality in regards of using a tool [16].

Besides, the operations for electronic books are limited to only tapping or swiping the screen and they lack the operational feel of using the book as a tool. In contrast to the usage of paper books with which readers read books finding the sensory reality such as feeling the texture of paper and picking pages to turn over them, the experience with electronic books is close to the sensation of viewing pictures rather than reading books because users operate them just by swiping screen to proceed pages.

Even though electronic books excel in the convenience and functionality, the sensory reality of using a book as a tool is lacked. As a result, users cannot have the physical interaction and operational sensations of the same quality as reading experience of paper books, and the reality decline of 'reading books' leads to the loss on the whole user experience for content [17].

To improve the "reading" experience of electronic books by representing the sensory reality of reading paper books, this study focuses on the interface and operation and incorporates the analog-like approaches into their design based on the AoD design theory [18]. Therefore, we would devise a new interface of the smartphone book application and propose the real-life experience book application with the design of the interaction reflecting the actual feel and physical movements of reading books (operating analog tools) [19].

3.2 Mechanism

Figure 1 shows the mechanism of the proposed application.

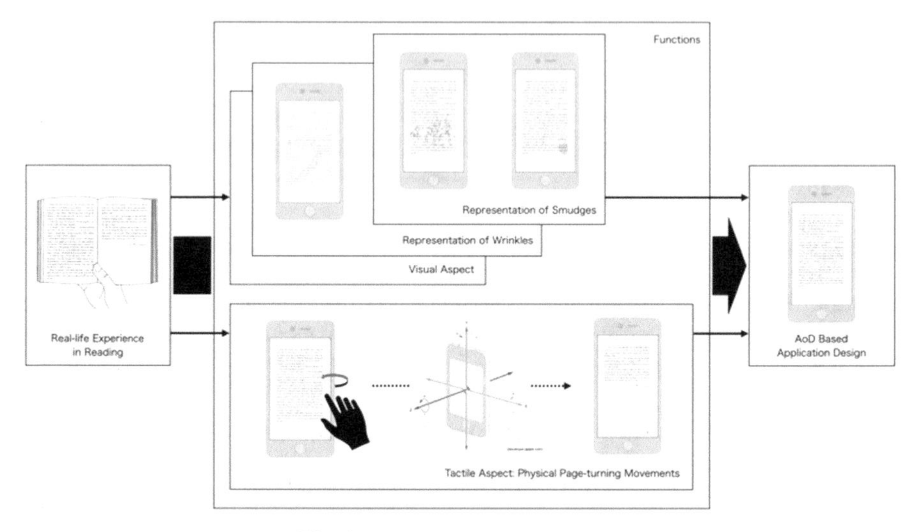

Fig. 1. Mechanism of application.

The application is for the smartphone and designed as a book application. In this paper, the application specifically targets iPhone by Apple. The application has the view function that supports the ordinary data styles such as text data, image file and PDF format. Application applies the function of UITextView to display text data and UIImageView to present image. The processing of PDF files is conducted in PDFKit, which is the software development kit equipped with Xcode to display/edit PDF files [20].

The interaction of application is constructed based on the that of paper books. First, to reflect the diverse situations of users into the application, the application incorporates "the content expression of changing condition" into the interface to represent the perceptual usability.

Regarding conventional electronic books, users operate the function of turning over pages by swiping screen. In this application, to reproduce the physical movements of turning over pages of the paper books, the system detects users' tilting movements for the operations on the device by the gyroscope and acceleration sensor to control page turn.

3.3 Visual Aspect of AoD – Design of Changing Perceptual Usability

Physical objects change through the long-term use and cannot always retain the constant condition. That change is recognized as perceptual usability and sometimes it functions as the appeal of the objects and enhances users' feeling of attachment. For example, leather products are presumed the use with deterioration and the discoloring of worn-out one would be praised as "good charm" [21]. Sometimes, people are even embarrassed to use a brand new one. This analog feature of "changing" allows people to experience the reality that "I am actually *using* it". In this way, people are fascinated by something with perceptual usability than the ones without perceptual usability.

Similarly, paper books change in condition through the repeated use. Figure 2 shows the same pages of the same books from two different conditions of the paper books that the author has.

Fig. 2. Comparison regarding perceptual usability.

The difference of perceptual usability is obvious. Especially, the lower shows the feature that how the owner reads it. Perceptual usability sometimes tends to be recognized as the deterioration and the falloff in quality, on the other hand, people feel more attached to the properly used old one than the brand new one. For example, the authors also have the feeling of attachment to the lower with perceptual usability rather than the upper, which is almost new because of the sensory reality that "I really have read this book".

On the other hand, digital content would never change in condition. Although the usage data would be uploaded and stored, even with the constant use, the appearance of the application design cannot provide the perceptual usability depending on the user's usage. Therefore, users cannot perceive neither the feeling of attachment nor the sensory experience by the application [22].

The book application proposed in this study makes the interface reflect the "changing" conditions of the book being used. It attempts to represent three changes that occur as a result of the use of paper books: “blurred or smeared letters”, “fingerprint smudges” and “wrinkled pages” [18]. Especially, “blurred or smeared letters” and “fingerprint smudges” are designed as natural “stains” which are naturally generated while reading books. On the other hand, “representation of wrinkles” is treated as feedback that comes from the interface in response to turning-page operations.

Figure 3 shows the designs of the stains: “blurred or smeared letters” and “fingerprint smudges”.

Fig. 3. Representation of “stains” as a sign of wear.

Blurred or smeared letters are the stains often seen in the paper books printed in ink. The problem of blurred or smeared letters has been ameliorated thanks to the advancement in printing techniques, and the quality of books has also improved. However, the representation of stains of blurred or smeared letters that occur when the book is touched produces a realistic sensation of reading printed books. Therefore, the application is designed to produce blurred or smeared letters when users turn over pages hard/swipe the screen or touch the same position repeatedly.

Fingerprints are designed to occur when the application is used for hours. These representations are based on the actual stains caused by sebum on fingers and sweat when users read books for a while.

It is important that the design of stain representation does not undermine the convenience the application offers to users. This application aims to provide an analog-like user experience with the representation of stains. In short, it needs to be designed to secure the convenience of a digital book application and to ensure a natural representation of stains without ruining the user experience.

Therefore, to meet readability and usability requirements, the application design will include a remedial function, which allows users to remove the stains by scraping the screen or stains to disappear after some time. The "changing" signs of wear in analog devices affect not only their appearance. There is a Japanese expression "to fit in hands", which describes an everyday phenomenon. It is used when the perceived shape, touch and weight of an object give you a desired sensation. It also implies that the user's habits result in the changes in the shape of the object over time. For example, shoes would re-form themselves to the shape of your feet as you wear them and get more comfortable. This means the changing feature of analog devices presents itself not only in the surface appearance but also in its form.

In the case of paper books, wrinkles are inevitable due to the material property of paper. Wrinkles affect the use of the books in addition to the surface appearance of them. If wrinkles occur, the page gets thicker than others, which makes it easier to open that particular page. Wrinkles reflect how users treat their books and the changes make the book "fit in (their) hands" (Fig. 4).

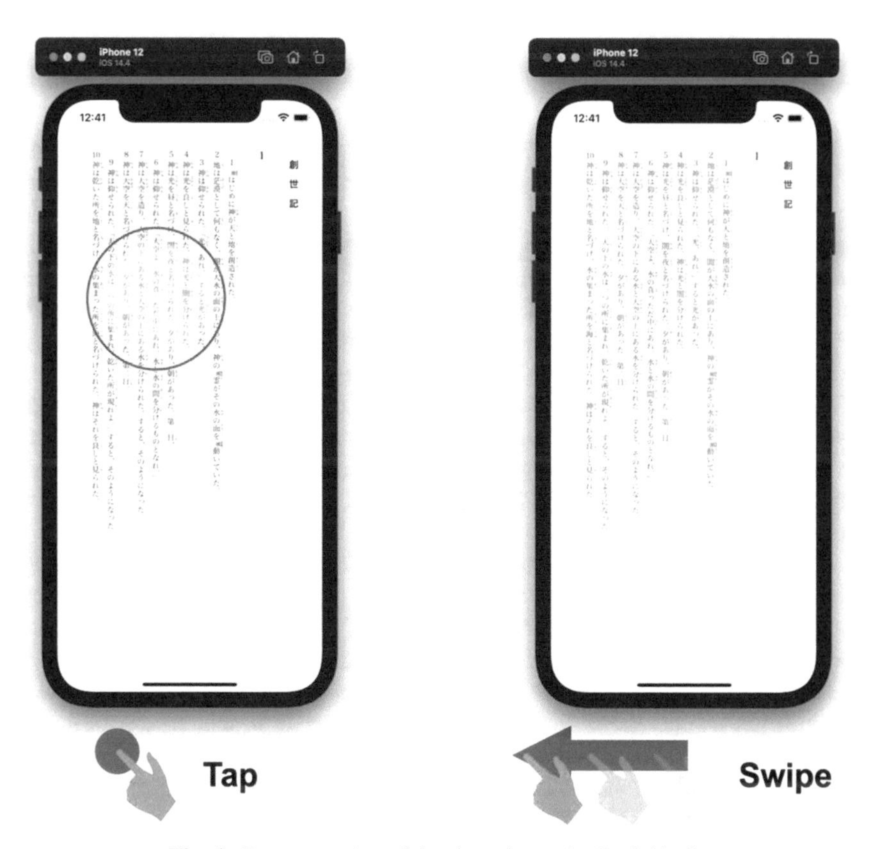

Fig. 4. Representation of the sign of wear by "wrinkles".

The application is designed to represent the wrinkles based on users' operations. It detects the intensity of the pressure of user's screen operation using pressure-sensitive 3D Touch function so that the condition of the wrinkles reflects the intensity [23]. First, the wrinkles generated by a tapping operation on the screen would appear as a small hollow. On the other hand, swiping would make the wrinkles appear as lines spread across a larger area. When the screen is operated, the stronger the pressure is, the greater the scale and the deeper the patterns of the wrinkles would be. Pages would get torn if the intensity exceeds the set value. Conversely, the wrinkles would be smaller in size and shallower if the pressure is weaker.

The ease of turning pages also depends on the condition of the wrinkles. We incorporate in the application the trait of paper books, which makes users to turn pages more easily where there are more wrinkles. When the application is launched, its first screen displays the page with the most wrinkles. We also devise an interaction in which users are directed to the page with the most wrinkles when they specify a particular page.

Just like in the design of stains, a remedial function is considered for the representation of wrinkles so that the convenience of the application is retained. Wrinkles would fade when users wipe the screen or spread them. If users reset the data, all the wrinkles would disappear and the book will return to "brand new" condition.

3.4 Tactile Aspect of AoD – Design of Realistic App User Experience Based on Physical 'PAGe-Turning' Movements

In addition to the representation of "changing" signs of wear in the interface, this study also attempts to represent readers' physical movements in the operation of the application. Despite the recent advancement in the ebook reader functionality, the basic operations are similar to those in the first-generation devices. Especially, the page-turning operation is performed only by one of two actions: tapping the end of the screen or swiping [24]. With thick books, this would take users a long time to turn pages, which can be quite inefficient [25]. To avoid this, users often use the seek bar to specify a page when skipping multiple pages (Fig. 5).

To jump to a particular page, users need to slide the button on the seek bar to select the page. Page-specifying operation on the seek bar lacks the intuitiveness readers enjoy with paper books when turning pages. Moreover, on average, opening the desired page on an ebook reader takes users 1.6 times as long and is 1.4 times as error-prone as with paper books [26].

It is easy to open a particular page in a paper book because the printed page numbers can be easily found. Also, holding and tilting a book to open a page are all done intuitively [27]. Therefore, this study represents the distinctive sensations of using real books and incorporates the physical movement of turning pages into the application (Fig. 6).

Current smartphones come with accelerometers and gyroscopes to precisely detect the 3-dimensional operations on devices. First, the speed at which readers tilt the book to open it tends to be higher when they turn more pages. At the same time, the angle at which the book is tilted tends to be greater. The application thus measures the changing values of device velocity on x, y and z axes with accelerometer, and detects the device's rotation and orientation with gyroscope.

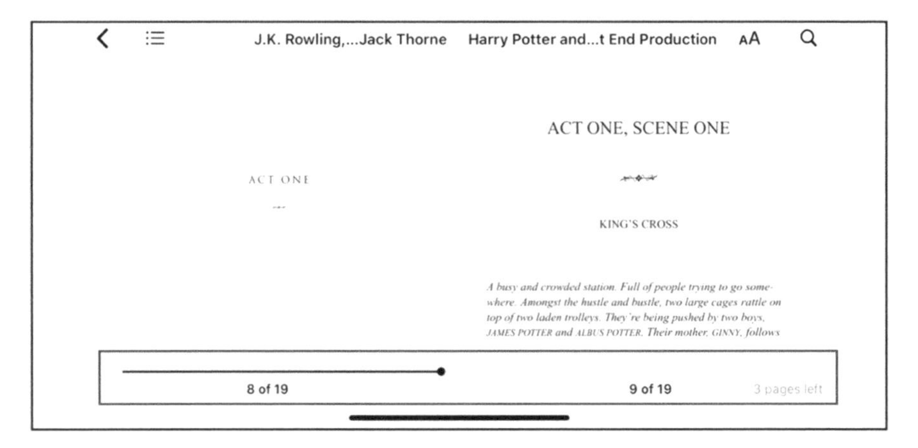

Fig. 5. Example of seek bar (iPhone iBooks).

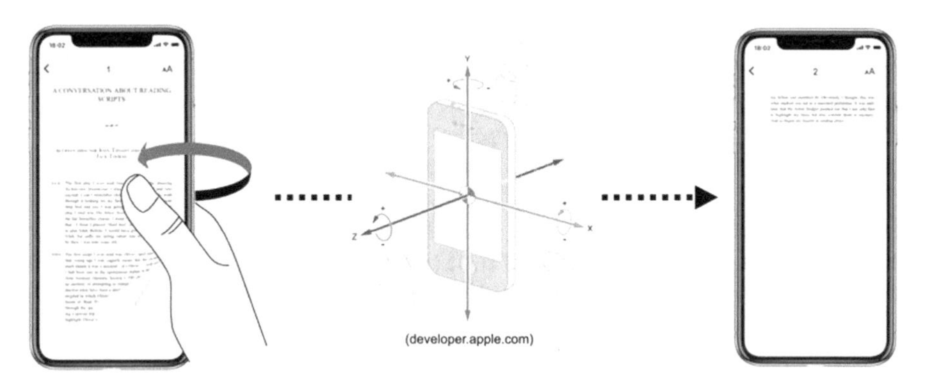

Fig. 6. Representation of physical page-turning movements.

The reported values will be incorporated in the application. The application then represents the physical sensations of turning pages of a real book with a tilted device in addition to the existing ebook operations - specifying pages by swiping and skipping pages on the seek bar [28].

In the page-turning movement representation, the application also incorporates the effect wrinkles have on perceived usability. Just like with paper books, the application will make it easier for users to open a page where there are more wrinkles when they do so by tilting the device. The design of the operation – "tilting the device to open pages" – represents the analog sensations of physical page-turning action on paper books, improving the efficiency in specifying the desired page on the app and merging the analog experience with digital application operations. This is also an attempt to digitally represent the whole experience of reading books by using the digital device as a paper book.

4 Conclusion

No matter how digital technologies advance, they could not meet the demand for analog devices. While digital technologies are for enhancing the effect of tools, analog devices epitomize the whole experience of using tools. In this paper we examined what digital and analog mean to today's lifestyle, taking the example of paper books and ebooks, and devised a digital method of applying "reality", which is a distinctive trait of analog devices.

This study advocates a new idea of design method that incorporates the element of real-life user experience into digital content, and considers its potential. It is also a proposal of a new use of content with a focus on "reality". This is different from conventional content uses, which depend on versatile digital devices for convenience and functionality. Especially from electronic books, users want reading experience, not the convenient functions of devices or applications. This paper designed a reality-experiencing book application to meet users' needs for analog-like reality based on AoD theory.

Future work should evaluate the potential effects of reality as an analog feature on the use of digital devices. While the research looks to prove the effects of reality and further improve the accuracy of the AoD theory, readjustment of proposed application is planned. This paper attempted to represent the sensory reality of using paper books from two aspects - interface and operations of the application. The universality of the method should also be examined in the future research.

The ultimate purpose of this study is to devise a next-generation digital computing method based on AoD theory. For this reason, the application example of reality described in this paper is just a basis of the theory development. In the future, we aim to examine the analog features from multiple aspects and apply them to digital content designs, and thereupon construct a mature theory system.

References

1. Appfigures blog. https://appfigures.com/resources/insights/ios-developers-ship-less-apps-for-first-time
2. Cambridge dictionary. https://dictionary.cambridge.org
3. Recording industry association of Japan. https://www.riaj.or.jp/f/data/annual/ar_all.html
4. All Japan magazine and book publisher's and editor's association. https://www.ajpea.or.jp/information/20190725/index.html
5. Yaguchi, H., Uemura, Y.: A study of the spread factors of e-book based on the social survey. J. Japan Soc. Publishing Stud. **42**(1), 123–142 (2012)
6. Fan, Z., Fujimoto, T.: Proposal of a design method to apply the analog features to digital media. In: International Proceedings of 2018 World Congress in Computer Science, Computer Engineering, and Applied Computing, pp. 3–9 (2018)
7. Liu, W.: Natural user interface - next mainstream product user interface. In: IEEE 11th International Conference on Computer-Aided Industrial Design & Conceptual Design, Yiwu, China, 17–19 November (2010)
8. Fan, Z., Fujimoto, T.: A method of implementing the sensation of operating analog tool on smartphone. In: 17th International Conference on Scientific Computing, Las Vegas, USA, 29 July – 1 August (2019)

9. Fan, Z., Fujimoto, T.: Proposal of a scheduling app utilizing time-preception-reality in analog clocks. In: 1st International Conference on Interaction Design and Digital Creation / Computing, Yonago, Japan, 8–13 July (2018)
10. Murota, H., Iinuma, T., Kubota, Y.: Simulation system for feel and weight of objects. In: Proceedings of Image Information and Television Engineers, pp. 188–189 (1998)
11. Nintendo labo website. https://labo.nintendo.com
12. Fujimoto, T.: Ideology of AoD: analog on digital -operating digitized objects and experiences with analog-like approach. In: 7th International Congress on Advanced Applied Informatics, Yonago, 8–13 July, Japan (2018)
13. Ministry of internal affairs and communications, Japan. White Paper, information and communications in Japan, ch. 1 (2020). https://www.soumu.go.jp/johotsusintokei/whitepaper/ja/r02/pdf/index.html
14. Ministry of internal affairs and communications, Japan. White Paper, information and communications in Japan, ch. 1. (2018). https://www.soumu.go.jp/johotsusintokei/whitepaper/ja/h30/pdf/index.html
15. Yaguchi, H.: A study of e-book as mobile media. J. Japan Soc. Publ. Stud. **40**(1), 45–62 (2012)
16. Fan, Z., Fujimoto, T.: Proposal of a digital book application that offers analog-like usability. In: Proceedings of the 15th IEEE Transdisciplinary-Oriented Workshop for Emerging Researchers, p. 26 (2018)
17. Yaguchi, H.: A comparative study of digital media and paper media from the viewpoint of text content legibility. J. Printing Sci. Technol. **49**(4), 252–257 (2012)
18. Fan, Z., Fujimoto, T.: Optimization of analog representation in digital media for book apps. Appl. Inf. Media Design **1**(1), 13–18 (2019)
19. Fujimoto, T.: Understandability design: what is 'information design'? J. Inf. Sci. Technol. Assoc. **65**(11), 450–456 (2015)
20. Apple developer website. https://developer.apple.com
21. Tsuchiya Kaban Inc. https://tsuchiya-kaban.jp/blogs/library/20200114
22. SAP business innovation update, "Considering digitization: the value of digitization and digital society". https://www.sapjp.com/blog/archives/13389
23. Apple developer, human interface guidelines. https://developer.apple.com/design/human-interface-guidelines/ios/user-interaction/3d-touch/
24. Kikuchi, J., Yaguchi, H., Kikuchi, M.: A study of evaluation of e-book interface. In: Proceedings of the 53th Japan Human Factors and Ergonomics Society, pp. 186–187 (2012)
25. Yaguchi, H.: A comparison study of the usability between electronic books and paper books about interface (Part 2). Japan. J. Ergonomics. **54**(supplement). 2G2–1 (2018). https://www.jstage.jst.go.jp/article/jje/54/Supplement/54_2G2-1/_pdf
26. Yaguchi, H.: A study of evaluation for usability on books. Japan. J. Ergonomics **50**(supplement), 264–265 (2014)
27. Yaguchi, H.: A comparison study of the usability between electronic books and paper books about user interface. Japan. J. Ergonomics **53**(supplement1), 208–209 (2017)
28. Apple developer, documentation. https://developer.apple.com/documentation/coremotion/getting_raw_gyroscope_events

The Modelling and Designing of Tourism Activity Locator Application

Ford Lumban Gaol[1(✉)], Tommy Wijaya[2], and Tokuro Matsuo[3]

[1] Computer Science, Binus University, Jakarta, Indonesia
fgaol@binus.edu
[2] Information System, Binus University, Jakarta, Indonesia
[3] Advanced Institute of Industrial Technology, Tokyo, Japan
matsuo@aiit.ac.jp

Abstract. Despite having so much variety of beauty in natural, cultural and historical heritage, Indonesia still has not maximized its tourism sector. The business problem identified was the lack of exposure on advertisement. The purpose of this thesis is to gain an in depth understanding of customer needs in order to bring more customers into tourism industry especially for tourism activity. By providing what the foreign tourist needs when visiting a country at the same time, this application exposes Indonesian providers and tourism activities to the global tourists.

1 Background and Idea

Indonesia has a great potential in its tourism sector by having a variety of beauty in natural, cultural and historical heritage. Based on the Indonesian Central Bureau of Statistics report, the number of visits from foreign tourists in the last 5 years shows a significant increase and keeps growing every year. In 2015, the number of visits from foreign tourists for one year reached 10.23 million visits compared with the previous year, the number increased by 8.43% [6].

Despite having so many resources in the tourism sector, Indonesia still has not maximized its tourism sector. Compared to its neighboring countries, Indonesia still has lower number of visits, as in 2015, Singapore was stated to reach 15 million of visits and Malaysia reached 27 million of visits [4].

The business problem identified was the lack of exposure on advertisement, especially for tourism activity. According to Shimp, 2010, exposure is as essential as its purpose for delivering messages through advertisement to potential customer or customer, which consists of brand image and business existence. Exposure means the occurrence of contact between advertiser and audience through magazines, radio advertisements, banners, internet, etc [10].

Seeing the increasing potential every year, it indicates that there is an opportunity in this sector. Therefore, the purpose of this business is to solve this business problem through the more effective marketing and promotions that meet the motives and interests of the tourists by creating a business plan for tourism activity locator [8]. The application will show the result of activity based on the keywords searched by the users and will

T. Matsuo et al. (Eds.): AIMD 2019, LNNS 677, pp. 14–21, 2023.
https://doi.org/10.1007/978-3-031-30769-0_2

recommend similar activity related to the keywords [3]. By listing all of the activities in Indonesia, especially for Bali, the authors wish to expose more and more tourism activities that Indonesia has to the global market. The purpose of the application is to provide what customer needs in travel preparations, such as information and location. By doing so, the application exposes the tourism activities to the customers.

2 Methodology

The methodology used in the preparation of this business plan is the descriptive analytical method. The data used consist of primary and secondary data. The primary data were obtained directly from field research that will be conducted during the business planning process [9]. The author will initiate the research by distributing questionnaire to its potential customers and analyze the data. Secondary data will refer to a literature study on relevant books, scientific articles and papers as the theoretical basis of this paper. The business planning process is based on the business planning process written by William Bygrave & Andrew Zacharakis, Entrepreneurship 2010 [5]. Customer analysis will be conducted by referring to a canvas used to have a better understanding on the customers written by Alexander Osterwalder, entitled Value Proposition Design and Business Model Generation [7].

3 Data Process Flow

The core function of this business is to advertise products from tourism activity providers to the market. Therefore, in order to effectively and precisely return the results from what user searched, the system must have an index of meta elements. The index acts as a temporary library of products with certain tags. Thus, it enables faster query matches and can get similar products in milliseconds. The data flow diagram can be seen in Fig. 1. Searching & Indexing Data Flow Diagram [6].

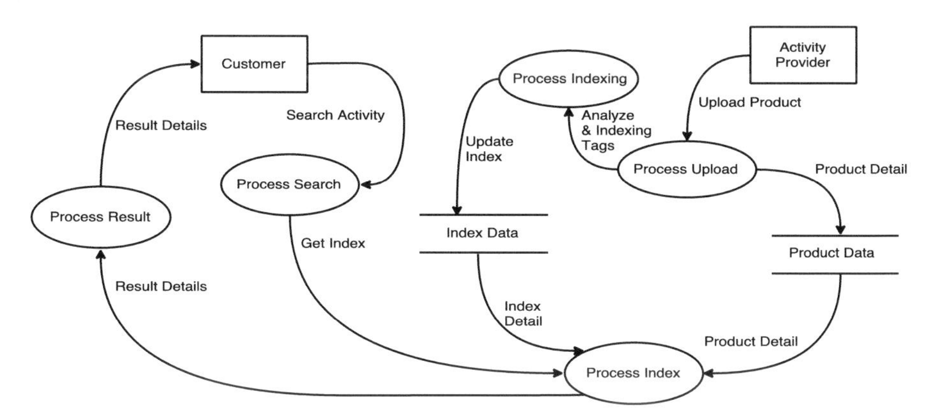

Fig. 1. Search & Indexing Data Flow Diagram.

There are two processes of data flow in Fig. 1. Searching & Indexing Data Flow Diagram [1]:

- Searching Data Flow sequences:

  1. Customer inputs keyword and searches through the application.
  2. The system processes the search by getting index details from index database and product details from the product database.
  3. The system returns the results to the customer. The results contain the name, price, category, location, and description of the product, and similar product details.

- Indexing Data Flow sequences:

  1. Activity Provider uploads the products they sell through the dashboard.
  2. The system processes uploading the product into product details database. Product details contain the name, price, category, location, and description of the product.
  3. The system proceeds to process indexing the product by analyzing the meta elements and updating the index database. Meta elements contain tags, category, product name, product types, location category, location tags.

4 Product Design

The product was designed to adjust the user needs. On Fig. 2 Find Activity Page, there are several components which have the same purpose that is to make the searching activity faster [2]. The find activity page is the homepage after the landing page. The landing page is when a user chooses to login or register. After logging in, the user will be directed to the find activity page. The find Activity page consists of 3 components of user interface design:

1. Search bar component

   Search is the most important component of this application. Therefore, the search bar is what user finds at the first glance. The search bar is positioned at the top of the page based on the human behavior in reading as it always starts from the top to the bottom. The customer journey starts when a user taps the search bar and starts inputting keyword.

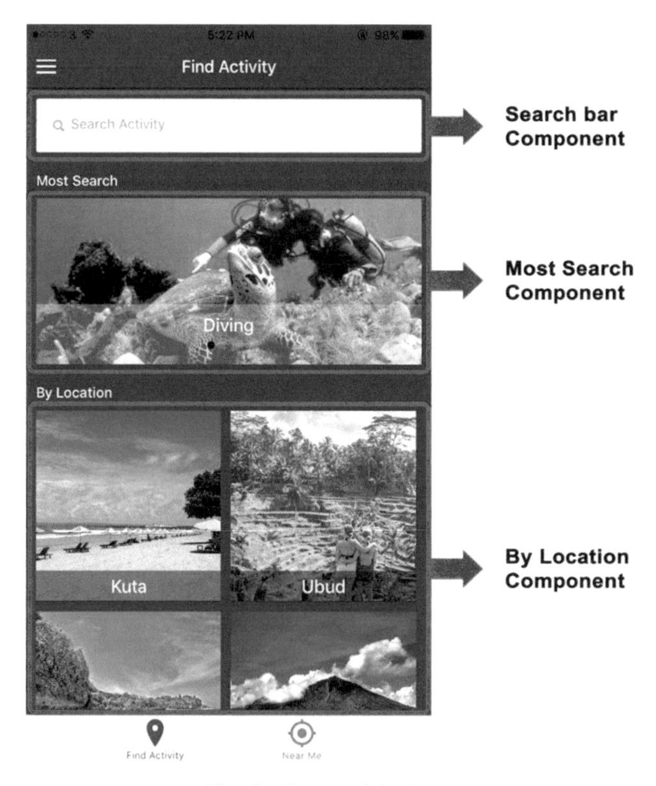

Fig. 2. Find Activity Page.

2. Most search component

   This component is a result of the indexing system on the most searched keywords and displays them through banners. The component can be swiped to left or right according to the number of banners available. This component is placed in the second order in the search bar to help the user to skip the process of keyword input. Each banner contains a keyword and will proceed to search the keyword tied to the banner when the user taps on it.
3. By location component

   This component contains places that are specially selected by the administrator of the system. The process of selecting the location is by researching the most popular locations for tourism activities and by calculating how many advertisements from the providers located in a specific location. This component is dynamically changing according to administrator's choice. A specific location with the most providers advertisement will be placed on top to boost the advertising features.

The second tab on the bottom menu is the Near Me page. This page displays a map and activity location inside the map (Fig. 3).

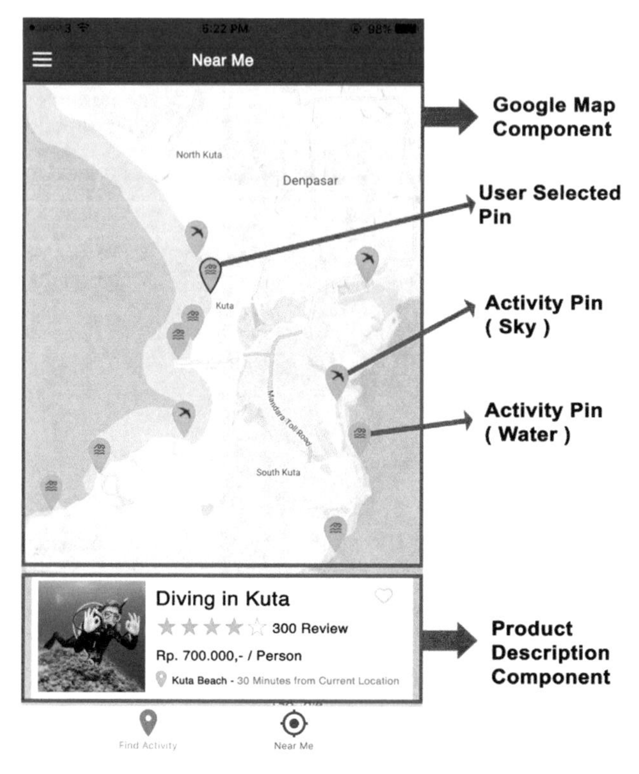

Fig. 3. Near Me Page.

The Near Me page consists of two groups of components:

1. Google Map Component.
   This component uses google map API to display map and user's location. On this component, the blue activity pin is the location of the provider. If the user taps on the pin, it will show the product description component. There are 4 types of category for the activity pin. Each category has an icon to show the type of category. The "bird" icon is for the sky-related activities category, the "sea" icon is for water sport-related activities category, the "mountain" icon is for hiking-related activities category, and the "human" icon is for land-related activities category.
2. Product Description Component.
   This component consists of the product details such as the image, price, ratings, and location of the product and how far it is from user's current location.

5 Technology

The internal application will have two separated layers, Frontend layer, and Backend layer as seen in Fig. 4. Technology Architecture.

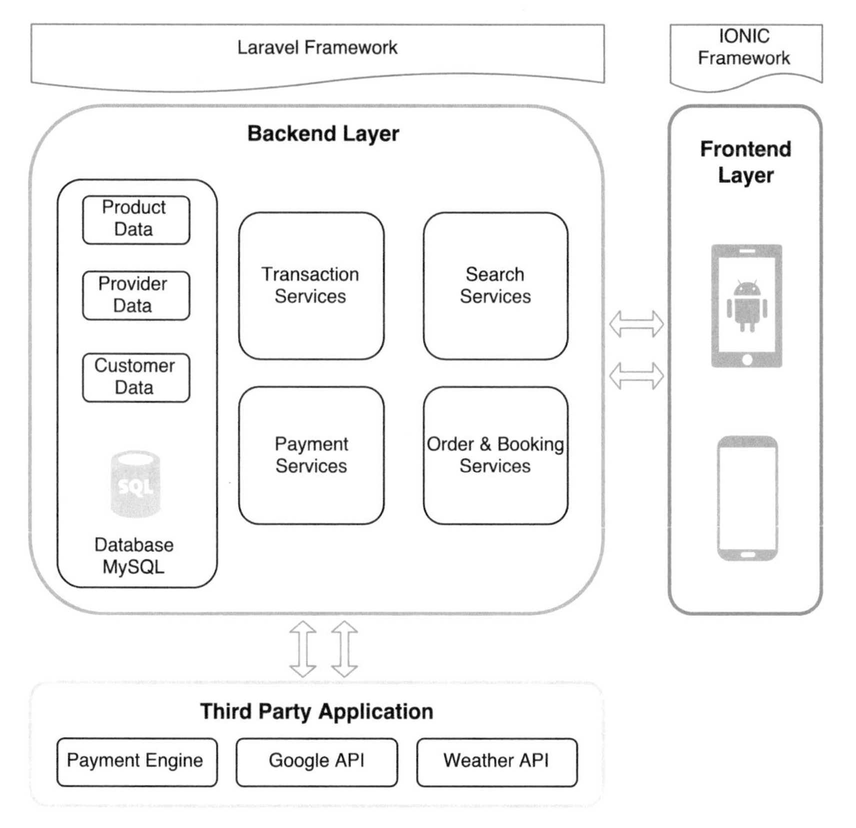

Fig. 4. Technology Architecture.

Frontend Layer

The Frontend layer is a layer that will be the user interface for the customer to interact with the system. The Frontend layer was built using IONIC Framework which supports both Android and IOS Platforms. IONIC Framework used Angular.js and HTML5 as the programming languages. The reason of using IONIC framework is that there are a lot of programmers understand those programming languages making it easier to hire a programmer in the future. The communications between the frontend layer and backend layer used HTTP request protocol with JSON as the language.

Backend Layer

The Backend layer will be built by using Laravel Framework which uses PHP 7.0 as programming language and MySQL as the database system. Backend layer consists of two groups of services:

1. Web Services

   Web services are the micro-services built to handle specific requests such as searching, booking and payment services. Web services receive a request from HTTP protocol via frontend interactions and process the data. The processing starts after receiving an HTTP request to book an activity, the web services then proceed to query the database to check for the availability and return to the frontend using the same protocol containing JSON data.

2. Database

   The database uses MySQL as the database language and its purpose is to store and process the data, such as the product data, the provider data, and the customer data.

Third-Party Application

The third-party application is an application from the third party that is integrated into the system using HTTP protocol. There are currently three applications that were integrated:

1. Payment Engine

   Payment engine is an engine that processes payment from customer and sends notification to the system. Payment engine has strong securities and already linked to many payment methods throughout Indonesia. The communication between this application and the system uses HTTP request protocol.
2. Google API

   Google API is an application that provides maps around the world. This application is used to pinpoint the provider's location and calculate the trip time. The communication between this application and the system uses HTTP request protocol.
3. Weather API

   Weather API is an application that provides weather information around a specific location. This application is used to forecast the weather around activities to see if it is possible to do the activity in that location in a selected time range. The communication between this application and the system also uses HTTP request protocol.

6 Conclusion

The presence of technology will help on exposing the content to a global market with the help of a good sense of purpose. Therefore, in order to validate this idea, authors suggest to publish the application and get real user feedback. To maximize the itinerary planning features, the business should have commodities and transportation as partners as well. This is because, at the end of the trip, the tourist will need a transportation and a place to rest before beginning the next adventure.

References

1. Andreansyah, R., Triyuniati, B.: The influence of promotion strategies towards foreign tourist's decision to visit. J. Sci. Res. Manag. **3**(6), 1–17 (2014)
2. Bygrave, W.D., Zacharakis, A.: Entrepreneurship, 2nd edn. Wiley (2010)

3. Carr, M., Verner, J.: Prototyping and software development approaches. City University of Hong Kong (1997)
4. Crisan, R.E., Berariu, C.: Advertising aspects of tourism. Acad. Sci. J. Geogr. Ser. **2**, 29–34 (2013)
5. Green, L.C.: 21st Century Business Entrepreneurship, 2nd edn. South-Western Cengage Learning (2011)
6. Lapian, S.Q.W., et al.: The influence of advertising and tourist attraction towards tour-ist's decision to visit Firdausi beach tourist attraction in north Minahasa district. **3**, 1079–1087 (2015)
7. Osterwalder, A., Pigneur, Y.: Business Model Generation. Wiley (2010)
8. Osterwalder, A., et al.: Value Proposition Design. Wiley (2014)
9. Pratomo, Y., Bakar, A.Z.B.A.: Utilization of human virtual intelligence framework in managing technopreneur knowledge, Bandung 574 (2007)
10. Shimp, A.T.: Advertising, Promotion, and other aspects of Integrated Marketing Communications, 8th Edition, South-Western Cengage Learning (2010)

Development of Drawing Software that Gives a Staple Drawing Feeling by the Difference in Sound

Yulana Watanabe(✉) and Takayuki Fujimoto

Graduate School of Information Sciences and Arts, Toyo University, Tokyo, Japan
{s4B102000049,fujimoto}@toyo.jp

Abstract. Drawing is one of the most familiar entertainments for us. From an early age, we take crayon and markers and draw lines freely, and as we grow up, we learn drawing at school. Not only that, we sometimes draw small icons on notebooks or sticky notes to express our emotions. It is deeply rooted in our life. With the spread of the Internet and the digitization of life, there is also a increase in the demand for drawing software, which is the digitization of drawing. There are various types of software, such as ones that can be enjoyed regardless of age, and those that have specialized functions for modeling or the use by illustration specialists. However, all of them pursued visual reproducibility and convenience. There is few software configured to bring out the sensation that a user get when he or she actually draw and user creativity.

In this study, it is assumed that drawing in reality is more efficient and the quality of the work is higher. Thereupon, we aimed to reproduce realistic drawing experience digitally. As the first step, 'drawing sound' that generates when each of three tools is used, was digitally implemented and an evaluation test was conducted. As a result, 90% of the participants made favorable comments to the drawing software of the proposal and the effectiveness of this study was proved.

Keywords: Drawing · Drawing Software · Sound · Real Feeling

1 Background

1.1 Digitization of Tools

With the distribution of many devices such as smartphones, our living environment has changed significantly. Tools such as cameras, notebooks, and watches that have enriched every day and made our life more convenient have been digitized. The contents of our bags were organized, and the bags have become lighter, and our mobility and convenience have been improved dramatically. We no longer do not have re-strictions by time and place.

But at the same time, it also means that the 'feelings' for each tool are diminishing. When we were little, we could not throw away the notebooks because the favorite animation/fictional characters were on the covers, even if we used up them so that we

T. Matsuo et al. (Eds.): AIMD 2019, LNNS 677, pp. 22–36, 2023.
https://doi.org/10.1007/978-3-031-30769-0_3

had no pages for writing any more. As an adult, wearing a well-known brand watch can mean a proof of high sophistication. In this way, tools were originally something to express the owners' feelings and personality, because people select and possess the ones they like from many options. In recent years, many tools have been converted to applications with uniformity and they are on smartphones and tablets. In other words, what we cared for with affinity is just data today.

If the content information is synchronized among various terminals, we use it without feelings of resistance.

On the other hand, while digitization is progressing today, many people still prefer to use conventional tools. The reason is thought to be that they have familiarity or special feeling toward the tools after the long-time use, in other words, they have personal attachment to the tools. For example, a ballpoint pen that someone gave as a gift at some time reminds us of the giver just by keeping it. Even if the ink of the pen runs out, we do not throw it away, instead, by replacing the refill, we try to keep it for the use as long as possible. 'Attachment' to tools can improve a user's concentration, quality of work, and work efficiency.

When using an analog tool that is actually touched and used in daily life, the characteristics of the tool are captured as a real feel. For example, depending on the type of ballpoint pen, smoothness, ink lumps, and ink bleeding upon drawing are regarded as the "flavor" of the tool. We select and use the tools that we like. It is difficult for us to have attachment to the tools that do not accompany with such "feeling of the use", and work efficiency also decreases. On the contrary, by using a tool with affinity, not only the improved work efficiency but also creativity may be brought by the chemistry of various factors of the tool. From these, we focused on the "feeling of the use" out of the factors that make us have feeling of using a tool.

1.2 Digitized Drawing Software

There are various types of drawing software products that have been in the market today. Since each of them has its own unique characteristic, people can select the software according to their needs. There are drawing software packages suitable for various purposes, and many of them have in common that they provide many choices for coloring and drawing tools, and that users can easily change/delete drawn lines and images. Drawing in reality begins with finding a proper place for drawing with the base material and drawing tools. If we use up the base material so that there is no space to draw on, it is necessary to look for a new one. The software completely solved this inconvenience. Not only that, the enlargement and reduction of the drawn object can be changed at the user's will, and detailed work and check for the whole image can be easily done. From these, it is much more efficient than actual drawing.

However, no matter how convenient they are, some people prefer traditional 'analog' to digital software. The reason is thought to be that from the digital software people cannot have the "real feel" that drawing in reality gives us. For example, even with the same act of actual 'drawing', regarding the user's obtained feeling, two cases are completely different: drawing with a brush and ink on Japanese paper, and drawing with a pencil on a notebook.

But digital software does not make this difference. Although there are differences in interfaces such as mouse and pen tab, there is no change in operations and use feelings. It is hard to get the feeling that you are drawing, because it is just 'moving' the device as operations. With the variety of drawing software products, our life is enriched in terms of convenience. On the other hand, it leads to giving up the feeling that we originally received.

1.3 Importance of 'Real Feel' in Digital

Today, digital terminals that are convenient to carry, such as smartphones and tablets are becoming widespread. We are also digitizing many kinds of tools and services. We are now able to use various tools such as calculators and timers that we have not always carried with us as one of the applications in the terminal. Despite of the im-proved convenience, the original charms of the tool have been lost as a result. One of them is the 'real feel' that is said to be a factor that constructs the experience of using tools.

Analog on Digital (AoD) theory proposed to incorporate analog charms to digital and to use digital in analog sort of way. In the research: "Research regarding the ex-tension of user sensory experience and applied design based on Analog on Digital theory" (2020), the authors digitally reproduced the analog physical sensation of using the tactile tool and extended it. It focused on four kinds of tools, such as watches and cameras, which are still popular as analog items despite tide of digitization. The usefulness was proved by conducting comparative experiments with three items: conventional analog items, existing applications, and applications devised based on the AoD concept. In addition, "Awareness Survey on the Demand for Reality" was carried out in the same study with 412 men and women in their 20s and 40s, and it shows that there were few negative responses to the reality and the demand for that. It was clarified that human beings generally need to feel reality.

2 Purpose

Recently drawing software products have various 'pen tools' as drawing tools. Not only brush and pencil, but also functions that imitate chalk and hard pens are included to improve the expressiveness of drawing. Some software packages cause subtle changes in handwriting in drawing reflecting the speed of drawing lines and the strength of the pressure toward the contact surface as the writing pressure. However, these functions are basically limited to the expressiveness of appearance.

There is almost no "real feel" when using tools in software. Although there are changes in the interface such as pen tablet, mouse, and touchpad, the user's operation itself is almost the same, and it may be difficult to have 'drawing feeling with the tool'. On the other hand, analog objects such as brushes and papers that are actually touched and holded are excellent in "real feel". For example, when using a pencil, we hear and feel sounds like 'knocking' and 'crisp'.

In the case of a brush, we dare to capture a very subtle sound, or even search for a quiet environment to capture the sound before drawing. This is a clear difference from digital, which is not restricted by time and place. Moreover, people can easily percept the differences not only in the drawing tools but also in the base materials. The sound is changed by the factors, base material: a general notebook made of paper, a special drawing paper, or a whiteboard (in case of markers). By changing the combination of 'drawing tool' (writing instrument) and 'base material' (drawing material), the sound is also changed.

When we draw in reality, we feel the 'reality' of using drawing tools on base materials by capturing the sound. However, drawing software has no function that pro-vides such a feeling.

Authors of this research believe that providing the "real feel" that can be obtained upon actual drawing, for digital drawing software, which has been recently developed in various ways, will lead to the recovery of users' affinity for tools. There are several factors that can be considered to enhance the feeling of the use of drawing tools, but in this paper, we specifically focused on 'sound'. We propose drawing software equipped with 'sounds' that gives users 'feeling of using drawing tool', motivate users, and improves work efficiency by enhancing user's affinity to tools.

3 Purpose Consideration

3.1 Reality by 'Sound'

It is natural for people to perceive something by sound in our daily life. For example, while walking, we hear the sound of a car approaching from behind, even if we do not actually see it. The importance of hearing is clear from the fact that it is forbid-den to ride a bicycle while wearing earphones or headphones and listening to music. Sound is also generated when we draw or work. In a quiet environment, the sound of drawing/writing something is naturally sensed. By capturing the 'drawing/writing sound' we know that someone is working or studying. At the test site, they imagine other's progress by the pencil sound of answering questions or the sound of turning a page, and the sounds arouse competitive spirit or sometimes they give people pressure. As for a computer, the sound of typing on the keyboard makes a user feel that the work is progressing somehow, and he or she feel like he/she is excellent. Readers may have experience of typing a keyboard at random without any content to be entered, just to feel like "I'm typing fast and it looks good" when you do not even know how to use a computer.

3.2 'Sound' that Gives a Reliable Feeling to Us

In our daily lives, we hear the sounds such as the sounds of cars, and the sounds that occur naturally. They can be referred to as "perceived sounds". Namely they are 'sounds for recognition and distinction'. But they are a little different from what gives us a reliable feeling. For example, in a fighting scene in a drama or animation, a hit-ting sound is generated, but it is different from the sound that is heard in actual fistfight. Furthermore, although few people have actually heard the sound of swords colliding with each other, a hard sound is always heard in scenes using swords. There-fore, it is often the case that people can judge a scene is a battle scene without looking at the screen. From this, it can be observed that when implementing sound in animation or content production, the actual sounds are not recorded and applied. In a famous example, the sound of a bird flapping is the edited sound of opening and closing an umbrella. Similarly, the sound of the waves is based on the sound of rolling beans in a colander. The sounds of the devices used in the SF content are all similar sounds. With careful thought, it is no wonder that the device is silent because of high-performance, but they all produce similar, quite inorganic beeping button sounds.

As these examples show, we do not feel uncomfortable with different sound from what we actually hear, and we get a sense of the 'fit' for the situation and objects.

Even for the situation/objects that we have never seen/heard, we can get a sense of the 'fit' from a sound that is close to our own imagination, without the actual sounds. The purpose of this research is to digitally provide the realistic drawing feeling by sound, it turns out that 'sound setting' is very important. In this paper, based on the free sound source that exists on the Internet as 'drawing sound', we developed and set sound that is close to real sound by referring to the actual sound's waveform.

4 Evaluation of Drawing Software with a Realistic Experience

In this research, we focused on the drawing feelings that we enjoyed when working in the real world as a factor that makes people have attachment to digital. We have developed drawing software for the purpose of providing attachment. There are various possible factors to provide a real feel, but we do not propose an implementation of all that can be thought at once. Factors will be provided in order, and it is assumed that they will be added in the process of improvement through subject experiments.

Regarding the functions to be implemented, we focused on the five sense stimuli for human perception. Specifically, this research features an auditory stimulus in drawing, in order to differentiate from the many existing software packages that dedicate to tactile stimuli and visual reproduction. That is why we implement 'sound'.

4.1 System Design

Depending on the selection and combination of the drawing tool and the base mate-rial, there are some differences in the sounds that are heard in reality. "A Drawing Software that Changes User's Realistic Experience by Sounds Generated by the Combination of Tools and Materials" (2019), clarified that there is a difference in sound when changing the combination of the drawing tool and the base material upon drawing in the same environment, by referring to the waveform for the examination. It also mentions the characteristics of the three drawing tools: brush, pencil, and marker.

In this research, based on the results of the precedent study, we implemented drawing sounds by processing free sound source on the Internet. The base material is assumed as high-quality paper to combine with sounds that was clarified in the waveform. This is because although there are differences in sound depending on the combined base material, the changes in sound by differences of drawing tools are more obvious. The most characteristic point regarding the brush, it is hard to hear the sound when using it in reality. Therefore, we implemented the brush sound at a clearly lower volume than other drawing tools.

Also, the feature was reproduced; there is no significant difference between the sound when people start to draw and the sound when the line is drawn continuously. It gives a rough impression compared to sounds by other drawing tools because it has a large friction surface with paper. Taking this into consideration, the sound was selected and edited.

The characteristic of the pencil sound is that at the beginning of drawing a line, there is a hard sound that hits a desk under the paper, and as the line continues to be drawn and it rubs against the paper. Reflecting this sound change, the click sounds at the beginning of a line and the sound when drawing is continued by dragging, are set to be different. The loudness of the sound was set to a 'standard' so that it can be heard even if there is some surrounding noise, while referring to the waveform of the actual recording of the brush sound. When conducting the evaluation experiment, the standard sound volume of the computer was set referring this pencil sound.

There are various types of tools defined as markers. In this research, we assumed a thick marker that has a wide contact surface with the base material and is also used to draw on cardboard. The sound characteristics are more tingly than the brush sound, and the overall sound is muffled. As a first try, we distinguished the sound when starting a line and the sound when continuing to draw a line as well as the pencil sound. However, from the results of the pre-test for evaluation, many respondents answered that without the distinction was better and we decided not to differentiate the quality of the sounds. Instead, we reflected the feature that the loudness gradually grows, little by little while drawing.

Considering that the marker sound is louder than sounds by the brush or the pencil, based on the waveforms, the sound was set to be louder than the pencil sound. The proposed drawing software aims to give users drawing feeling by sounds. Therefore, other factors such as visual expression of drawing were not changed, and the thick-ness and darkness of the drawn lines were unified. The interface was designed to draw using the trackpad or mouse that participants are familiar with in daily life.

4.2 Evaluation Experiment

We asked 38 participants to use the prototype of the proposed drawing software, and conducted a questionnaire regarding the feeling of the use, the presence or absence of work efficiency, and the difference when compared to the existing software. The participants are 38 university students who casually use computers and digital devices and also have experience in using existing drawing software. In the experiment, at first, participants were requested to draw in a situation where they could hear only the click sound without the sound of the writing instrument. The request had two purposes: Users recognize the drawing in the silent state and get used to the operation. After that, they were asked to draw with the sound of each writing tool. Participants were free to draw what they wanted, but we asked them to include both drawings and characters such as pictures and lines. And every time they finished using each drawing writing tool, we asked them to answer a questionnaire about it.

There are two questions; one is 'which is better, with or without sound?', with three options including "Neither.", and the other is a five-scale evaluation regarding the enhancement of the drawing feeling. In addition, we asked them to answer their impressions and points for improvement in a free description. Figure 1 shows a drawing screen of the prototype drawing software. The screen is the same for each drawing tool because the expressiveness of the appearance does not change regardless of the use of different drawing tool by default.

Fig. 1. Drawing screen of prototyped drawing software

4.3 Evaluation

The questions were narrowed down to the two questions described below and the questionnaire was conducted. The experiment was conducted with a small number of people so that they could ask for help in filling out the questionnaire, and also, they could tell if they had any questions or comments. In this way, we tried not to miss any frank questions or impressions.

1. Was the 'drawing feeling' enhanced when you drew with sound?

   [Extremely well/Very well/Moderately well/Slightly well/Not at all].

2. Which enhances the feeling of drawing more, with or without sound?

   [With sound, Without sound, Neither].

Pencil
The questions were narrowed down to the two questions described below and the questionnaire was conducted. The experiment was conducted with a small number of people so that they could ask for help in filling out the questionnaire, and also, they could tell if

they had any questions or comments. In this way, we tried not to miss any frank questions or impressions.

Regarding the results of the questionnaire carried out after drawing with the sound of a pencil, there are more favorable answers than other drawing tools. As a result we were able to obtain answers from 23 people that the drawing feeling was enhanced "extremely well". This is more than 60% of the total. In addition, 13 people chose the positive answer of "Very well", and including this, 95% of the participants answered that the drawing feeling was more enhanced with sound. In addition, 37 out of 38 participants answered that they had a drawing feeling. On the other hand, there were some participants who answered "Neither" as to whether or not there was drawing feeling due to the presence of sound. The remarkable comment is, "It was very pencil-like compared to other drawing tools. However, it might have been recognized as the sound of a pencil because I drew after knowing that it was a pencil., If this was called the sound of chalk, I might thought it felt like a chalk."

Regarding drawing with a pencil, there were many good reactions from the free de-scription column. In addition, during the experiment before the questionnaire, there was a comment: "it feels like writing with an actual pencil, and it's fun!" Many people answered that the hard sound, which generates when the pencil contacts the desk at the beginning of drawing, had led to the drawing feeling, rather than the "smooth" continuous sound generated while drawing lines. After the experiment and the questionnaire, while exchanging conversations with the participants, it was thought that the sound of the pencil will be more realistic by expressing the roundness of the core. As a characteristic of a pencil, when it is just after being sharpened, the core of the pencil is sharp, and it makes a crisp sound.

In the experiment of "A Drawing Software that Changes User's Realistic Experience by Sounds Generated by the Combination of Tools and Materials" (2019), the difference between the sharpened state of the pencil and the rounded state after using it to some extent was specified. However, in this prototype, since the pencil sound was implemented assuming a sharp state, it does not change even if drawing is continued for a long time. We were able to get comments on this point from the participants. According to them, the "non-smooth sliding" sound of the pencil after continuous use gives a user the recognition that "it is high time to sharpen the pencil", and it leads to the recognition that "I have used this pencil so much." From these, it was possible to confirm the importance of the sound presence for drawing.

The answers to questions 1 and 2 in the questionnaire regarding a pencil, are shown in Figs. 2 and 3 below.

Brush

Regarding the drawing feeling enhancement upon drawing with sounds, there were 17 people who answered, "Very well" / "Well" for each. In total, 34 people gave positive answers. Regarding the comparison of sound presence or absence, 34 responded that "With sound" was more realistic, while 2 selected "Neither", and 2 responded that "Without sound" was better. From the free-description answers, it became clear that t the brush sound itself was not well known, regardless of whether analog or digital.

Upon the prototype application, the brush sound was set as very slight sound based on the comparison with other drawing tools while referring to the waveform of the

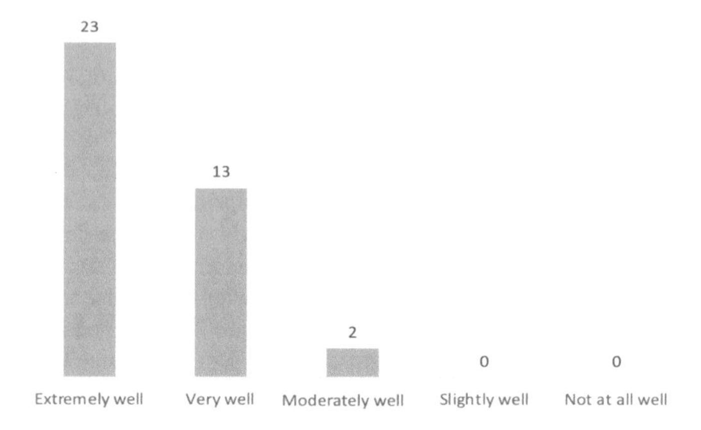

Fig. 2. Answer to question 1 when using pencil.

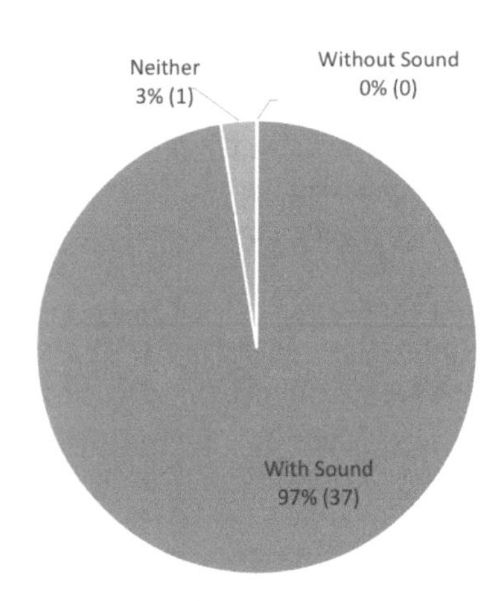

Fig. 3. Answer to question 2 when using pencil.

actual sound. In response to this setting, a participant said, "I got a feeling close to that of actual drawing because I had to concentrate to listen to the sound working in a quiet environment as the required style. In this way, the drawing feeling was brought, and it is not just because of the sound function of the drawing software. Another participant selected "Neither" saying, "I can't remember the sound of using a brush when drawing in reality, and I can't really imagine it. Therefore, it was difficult to recognize that the implemented sound was the brush sound."

In addition, there was a participant who selected "Neither" because "I knew I heard something, but it was difficult to judge that it was the brush sound." (They were not allowed to change the experimental environment) Since what to draw was arbitrary, there were some comments such as "(What I have drawn seems to make a more soft, ink-spreading, and blur sound."

To examine this comment, after the experiment, we actually put a brush tip over the paper and moved it to draw a line slowly while being conscious of spreading, and just drew a line as usual. As a matter of course, there was a difference in length of the sound. However, the volume of the sound was almost the same, and it was not possible to identify the "spreading sound" even while working with the brush. It is possible that people receive the visual stimulus and reflect it to sound recognition unknowingly. We developed the software supposing the same condition that is not affected by vision. However, it was found that it is necessary to pay attention to the fact that other factors such as the working environment and visual expression are greatly involved in the expectation for the brush sound, instead of the actual sound.

The answers to questions 4 and 5 of the questionnaire regarding a brush are shown in Figs. 4 and 5 below.

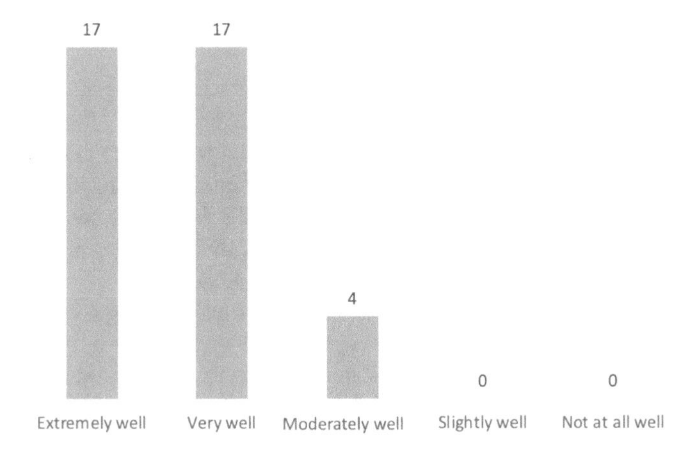

Fig. 4. Answer to question 1 when using brush.

Marker

Regarding enhancement of drawing feeling, 33 participants gave positive responses for drawing with the sound, but 4 answered "Moderately well" and one selected "Slightly well". As to the questions about the presence/absence of sound, 32 people answered, "With sound is better", and majority of responses were favorable. However, there were 4 participants who selected "Neither", and 2 answered "without sound is better" (as to other drawing tools, no one answered "without sound is better." As for the question about work efficiency, the comment of the participant who answered, "Slightly well" was "The sound made me feel like I was drawing, but I don't like the marker sound anyway, so it was difficult to improve the work efficiency.".

Out of the participants who answered "Neither" to Question 2, there were 3 people who thought that it was a question about the preference of with or without the sound.

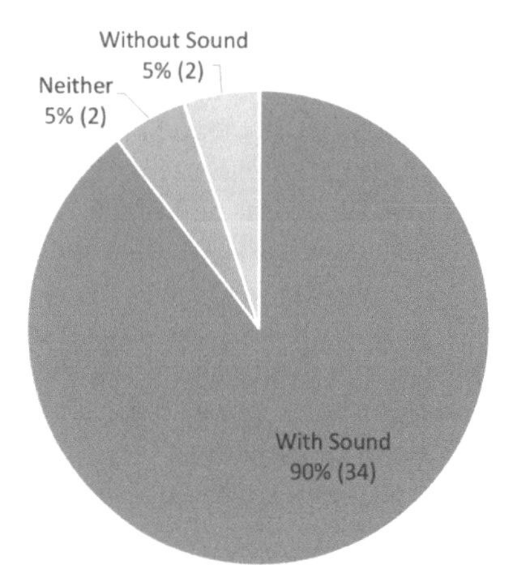

Fig. 5. Answer to question 2 when using brush.

They did not understand it was a question about presence/absence of drawing feeling. Same happened to the one, who selected "without sound is better." From them, we got the answer "with sound is better " from them based on the correct understanding that the question focuses on the presence /absence of the drawing feeling,

From the picking frank comments and free descriptions, it is considered that the most characteristic feature of the marker sound is that its impression is associated with each user's preference when compared with other drawing tools..

There are various types of markers compared to other drawing tools. In the experiment, the following description was conveyed to the participants as the image of the marker that was assumed to generate the sound; the pen tip is thick and used on not only paper but also cardboard as a base material. However, it was difficult to share the exact same image, and some participants in the experiment said that the sound was different from what they had expected. Many participants were disliked the sound of the marker besides presence/absence of drawing feeling or the work efficiency. As a direct question, when the authors asked the participants "Do you actually want to use the marker with the sound?", most of the responses were negative, while they gave positive responses for other drawing tools.

The answers to questions 1 and 2 of questionnaire regarding a marker are shown in Figs. 6 and 7 below.

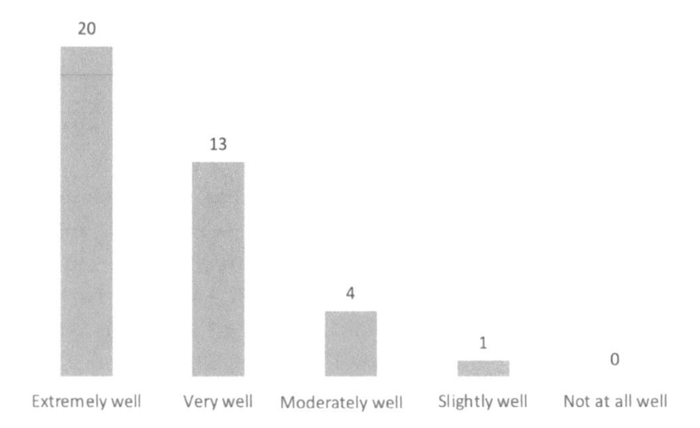

Fig. 6. Answer to question 1 when using marker.

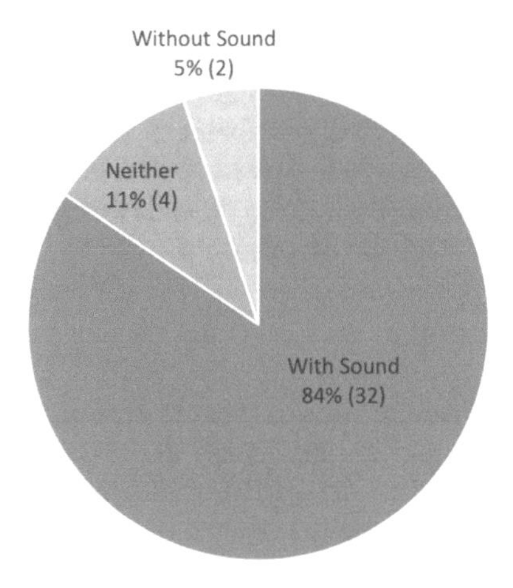

Fig. 7. Answer to question 2 when using marker.

5 Conclusions and Future Work

5.1 Conclusion

In recent years, with the digitalization, our life has become more convenient, but the feelings of use and attachment that we have traditionally had for various things are fading. Drawing is one of them. When we were children, we used crayons and chalk to draw rough lines of our choice or for coloring as entertainment. When we grew up, we learned drawing such as 'sketches' in school classes. However, in recent years, digital drawing has made it easier to use thin lines such as those written with a pencil from the start of drawing experience. Some children does not have experience of drawing lines by themselves, because they just have the experience with coloring books by default. Even the coloring book is colored by touch on digital, so they may not know how to draw moving their arms.

It is certain that anyone can freely draw 'acceptable' pictures with the development of tools, but it also means that no matter who draws, only similar works can be produced. This can be rephrased as each user's expression of creativity is lost.

It is thought that the affinity to the tools and the feeling of 'the fit' to the work are generated by the stimuli from sight, touch, smell, taste, and touch. In this research, we developed and examined drawing software equipped with 'sound', in order to digitally reproduce the reliable feeling that we have when actually drawing, based on the previous research. The drawing with the three types of writing tools was compared with the conventional drawing, and the evaluation was carried out by a questionnaire. As a result, more than 80% of the participants in the experiment said that it was good. Compared to conventional software, it was easy to arouse the feeling of realistic drawing experience. On the other hand, in some cases, the sound of strangeness, and the difference between the implemented sound and the sound in the user's expectation/imagination, caused confusion and even caused discomfort against the purpose.

Majority of the impressions about pencils were particularly positive. Regarding a brush and a marker, it was pointed out that they could not be understood because they had not known the brush sound when writing in reality, and the type of marker was different from what they had expected with their expectation.

From this, as discussed in Sect. 3.2, we reconfirmed that the "perceived sounds" that are actually heard when using tools, and the "sounds that gives reliable feeling of 'the fit'" are different through the subject experiments.

5.2 Future Work

This research aims to reaffirm the attractiveness of analog, which is being lost due to recent digitization, and to equip it with digital. The drawing software proposed and developed in this paper is the first step of the whole project. Among the sensory factors that can be obtained when actually drawing, this paper focused on 'sound' based on previous research and implemented it. However, from hypotheses and experimental results, it became clear that there is a difference between the sound generated when drawing in analog and the "sound of the 'fit'" that is easily connected to the reliable feeling of drawing. For further study, the first task is to clarify such 'perceived sounds' and 'sounds that easily fits to user's imagination/expectation.' Along with this, we will proceed with editing and development of the system so that we can implement sounds that give users a reliable feeling closer to that acquired upon actual drawing experience. The attractiveness of 'analog' regarding drawing cannot be re-produced only by the 'sound'. It is considered that the tactile element also constitutes a large portion because drawing accompanies with one's physical movements. Be-sides, the inconvenience of analog, such as tools' age-related changes and the need of sharpening a pencil or adding ink, may arouse users the feeling of affinity. We would like to examine various factors and advance the development of drawing software that has charms in analog sort of way.

References

1. Tanaka, Y., Nogai, T., Munetsuna, S.: Asymmetrical audiovisual perception in temporal order judgment. J. Inst. Image Inf. Telev. Eng. **32.48**, 21–26 (2008)
2. Okamura, T.: Importance of visual, auditory, and tactile senses in sensory integration. Int. J. Affect. Eng. **11**(3), 503–507 (2012)
3. Yokokura, Y., Yokokura, Y.: Stereo production technology, stereo CM production, the actual creation of commercial sound effects [in Japanese]. J. Inst. Telev. Eng. Jpn. **34**(4), 317–319 (1980)
4. Fujimoto, K.: Charm of analog record (4) Charm of single board. Impact Publishing, Ambos Mundos (4), 96–98 (2000)
5. Ohisa, N., Yoshida, K., Yanbe, T., Kaku, M.: Effect of autonomic nervous system activity while listening to music. Auton. Nerv. Syst. **42**(3), 265–269 (2005)
6. Deguchi, S., Furuta, N.: A study on the generation of sounds and figures using LCD tablet. IPSJ SIG Tech. Rep. **115**(2) (2005)
7. Sugasawa, E.: Image editing software for creative use (5) Make picture letters with paint tools. Nikkei PC (501), 133–136 (2006)
8. Furukawa, K.: Development of pocket painting tool using USB memory. Res. Rep. JET Conf. **2006**(6), 65–68 (2006)
9. Kasai, S., Goto, Y., Murayama, Y.: e-gakki : this system convert drawing information to sound. In: The 69th National Convention of IPSJ, 2007, no. 1, pp. 487–488 (2007)
10. Kimura, A., Omachi, H., Shibata, F., Tamura, H.: A study for presentation of tactile sensation with sound feedback to touch sensor. Inf. Process. Soc. Jpn. (IPSJ) Hum.-Comput. Interact. (HCI), 2007(68(2007-HCI-124)), 9–16 (2007)
11. Katsutani, Y., Nagai, Y., Morita, J.: Influence of music on drawing. Bull. Japan. Soc. Sci. Des. **59**, 59 (2012)
12. Fujimoto, T.: Toward information design 3.0: the information design for 'com-municate.' Build. Maintenance Manage. **34**, 42–46 (2013)
13. Fujimoto, T.: Understandability design : what is 'information design'? J. Inf. Sci. Technol. Assoc. **65**, 450–456 (2015)
14. Kato, S., Mizuno, S.: Improvement of the analyzing method for drawing and sound generation system "RAKUGACKY." Digital Contents Creation **5**, 11–19 (2017)
15. Fan, Z., Fujimoto, T.: Implementation and evaluation of document production support system "DLO-Editor" with obsession mechanism". In: 25th International Conference on Systems Engineering (2017)
16. Fan, Z., Fujimoto, T.: Implementation of document production support system with obsession mechanism. In: Lee, R. (ed.) CSII 2017. SCI, vol. 726, pp. 51–64. Springer, Cham (2018). https://doi.org/10.1007/978-3-319-63618-4_5
17. Fujimoto, T.: Ideology of AoD: analog on digital-operating digitized objects and experiences with analog-like approach. In: 1st International Conference on Interaction Design and Digital Creation / Computing (IDDC 2018) (2018)
18. Tanaka, Y., Fujimoto, T.: A design of application to turn a smartphone into a computer mouse and possibility of preventing from being copied. In: 1st International Conference on Interaction Design and Digital Creation / Computing (IDDC 2018) (2018)
19. Fan, Z., Fujimoto, T.: Proposal of a design method to apply the analog features to digital media. In: The 2nd International Conference on Applied Cognitive Computing (2018)
20. Fan, Z., Fujimoto, T.: Proposal of a scheduling app utilizing time-perception-reality in analog clocks. In: 1st International Conference on Interaction Design and Digital Creation / Computing (IDDC 2018) (2018)

21. Fan, Z., Fujimoto, T.: Proposal of a digital book application that offers analog-like usability. In: The 15th IEEE Transdisciplinary- Oriented Workshop for Emerging Researchers, p. 26 (2018)
22. Fan, Z., Fujimoto, T.: Method to control children's smartphone use based on the motif of analog fuel system. In: ICSTR Bangkok – International Conference on Science & Technology Research (2018)

The Impact of Video Conference Application to College Students in Online Learning Activities

Givbrela Lostawika[1](✉), Devi Siti Azzahara[1], Marven Immanuel Christianto[1], Nunik Afriliana[1], Tokuro Matsuo[2], and Ford Lumban Gaol[1]

[1] Bina Nusantara University, Jakarta, Indonesia
givbrela.lostawika@binus.ac.id
[2] Advanced Institute of Industrial Technology, Tokyo, Japan

Abstract. Covid-19 pandemic has brought many changes in several aspect such as health, economy and education. The pandemic condition enforced the society to change their usual habits and activities to the new normal to deal with this outbreak situation. Education sector is one of many sectors influenced by the pandemic. Many educational institutions transformed their traditional face-to-face teaching method to distance learning, facilitated by some online teaching platforms such as Zoom Meting. This research was aimed to evaluate the impact of Zoom Meeting utilization on students learning result during their online learning period. It combines a descriptive technique with a quantitative approach. Research samples were collected from 32 students from 15 different institutions. This study found that many students join the online learning through the Zoom Meeting application and shows that the use of video conferencing media increases the time flexibility. Online learning was run well. However, not all the material can be conveyed properly, hence students have to elaborate the learning material by themselves. Some problems were also frequently occur in the utilization of zoom meetings such as the internet connectivity and the time restrictions.

Keywords: Online Learning · College · Video Conferencing · Zoom Meeting

1 Introduction

Corona virus pandemic has influenced the health, economy, and education. Numerous changes have happened, most notably in the area of education, where the traditional face-to-face teaching method has been replaced with distance learning through online media in order to halt the spread of the current COVID-19pandemic. Education system is suspectable to this pandemic dangers [1–3]. Governments have enacted many rules, one of which governs how distant education is treated. The University may continue to provide remote learning through a variety of channels. Universities often utilize systems such as Zoom Meetings, Google Classroom, Webex, and Ms. Teams [4].

T. Matsuo et al. (Eds.): AIMD 2019, LNNS 677, pp. 37–47, 2023.
https://doi.org/10.1007/978-3-031-30769-0_4

Zoom Meetings is a software program developed and launched in January 2011 by Eric Yuan. Zoom is available through the Zoom website and on a variety of operating systems, including Mac OS, Windows, Linux, iOS, and Android. Zoom meetings is a video-based communication tool [5]. The ZOOM Meetings program was created to enable distant conferences via the use of video conferencing, online meetings, and chat, as well as mobile collaboration [6, 7]. This program is frequently used as a means for long-distance communication between friends, coworkers, and family members.

Numerous mobile devices, such as telephones, desktop computers, and space systems, so that users may continue to conduct learning communication activities correctly, comfortably, and flexibly [8].

This research examined the effectiveness of online learning tools during the COVID-19 epidemic. It compares the effectiveness of Zoom to more traditional methods such as face to face instructional design. The following Research Questions (RQs) were constructed:

RQ1: Does implementing online learning activities via Zoom meetings have a good impact?
RQ2: Dose it effective for college students?
RQ3: What are substantial differences between online learning through Zoom Meetings and on-campus face-to-face learning?
RQ4: Which features do users prefer when utilizing zoom for online learning?
RQ5: What are the problems experienced by the students while utilizing Zoom meetings?

2 Literature Review

Online learning is structured learning with the goal of using an electronic or computer system to assist and support the learning process. It is a term refers to an educational system or idea that makes use of information technology to facilitate the learning process. The learning system is used to facilitate the teaching and learning process between instructors and students without the need a direct face-to-face contact [9, 10].

There are many primary features of online education. When referring to the first feature, which literally translates as electronic or online learning, it can be stated that this approach necessitates the use of electronic and digital technological services. The second distinguishing feature is the instructional materials. It is often in the form of self-contained instructional materials in digital format in online learning. The instructional material is then saved in a computing system, which enables instructors and trainees to access it from anywhere and at any time. Online learning features enable the use of learning schedules, the development of curriculum, and the administration of systems that are accessible at any time through computer networks [10–12].

Numerous parties may feel the effect and advantages of online learning, one of which is students. It enables students to do remote learning activities more easily during the COVID-19 epidemic. They may get instruction from instructors through video conferencing software [6] such as ZOOM Meetings. It is very beneficial for students to be able to perform activities such as asking and answering questions, completing group tasks, and teaching and being taught simply via ZOOM Meetings. Additionally, communication through the ZOOM Meetings is a very simple activity and eliminate the distance barrier for every students [7, 13, 14]. According to [8] lectures given through Zoom Meeting are insufficiently effective since difficulties with the network or internet connection often occur for students who do not utilize Wi-Fi. It is lowering the quality of the learning. However, the benefits of Zoom Meeting's usage are deemed practical and efficient for students, since contact between students and lecturer is much simpler.

The variables affecting online learning media research are examined in depth, with a particular emphasis on the connection between teaching behavior, students and lecturers, learning culture, and the effect of the Covid-19 epidemic. The student engagement is also an interesting aspect to be considered when it comes to create successful online learning activities [15].

3 Material and Method

This study was conducted in 2020. We used a descriptive and quantitative research technique. The data were gathered by distributing questionnaires. We utilized this google form since it is a more efficient and comprehensive way of distributing questionnaires to participants. The study was consisted of three stages: (1) creating questions to be distributed to students using Google Forms, (2) disseminating questions online via social media in the form of WhatsApp, and (3) collecting and filtering the data collected from students for subsequent analysis.

The questionnaires were filled by 32 respondents from 15 different universities participating in online education during the Covid-19 pandemic. The respondent's characteristics are classified according to their university and semester of study (Fig. 5).

Table 1. Respondent's Current Semester.

Semester	Student	Percentage (%)
1	4	9,4%
3	22	68,8%
5	3	9,4%
6	1	3,1%

Table 2. Respondent's Affiliation.

Universities	Number of Student
Bina Nusantara University	15
Diponegoro University	1
University of Indonesia	1
Brawijaya University	2
Padjadjaran University	2
UKRIDA	1
Sekolah Tinggi Penerbangan Indonesia	1
Gajah Mada University	1
UBD	1
Pelita Harapan University	1
Prasetiya Mulya University	1
Sanata Dharma University	1
Mercu Buana University	1
PKN STAN	1
UIN Jakarta	1

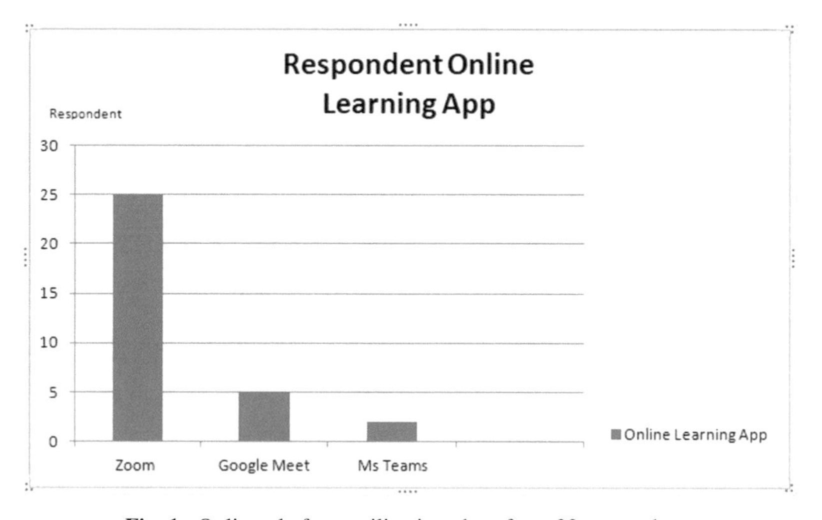

Fig. 1. Online platform utilization chart from 32 respondents

Table 3. Online Platform Utilizations.

Application	Total Respondent Number	Percentage
Zoom	25	78,1%
Google Meet	5	15,6%
Microsoft Teams	2	6,3%

Table 4. Respondent's Current Semester.

Reasons	Total Respondent Number	Percentage
University Recommendation	31	96,9%
Convenience	1	3,1%

Table 5. Reason for Using Video Conference Platform.

Frequency	Percentage
1–2 days per week	6,2%
3–4 days per week	43,8%
All day	50%

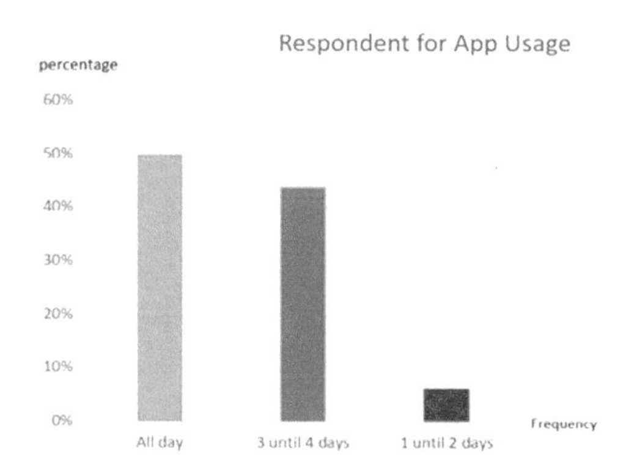

Fig. 2. Online platform utilization chart from 32 respondents

Table 6. Perceive of Effectiveness.

Effective	Total Respondent Number	Percentage
Yes	18	56,3%
No	14	43,8%

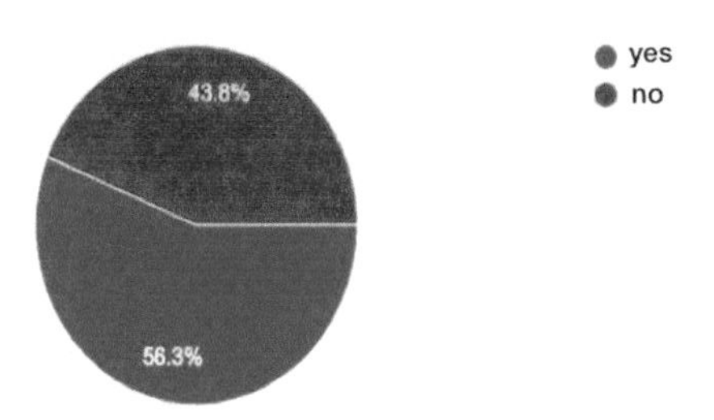

Fig. 3. Perceive of Effectiveness

Table 7. Perceive of Flexibility.

Versatile/Flexible	Total Respondent Number	Percentage
Yes	28	87,5%
No	4	12,5%

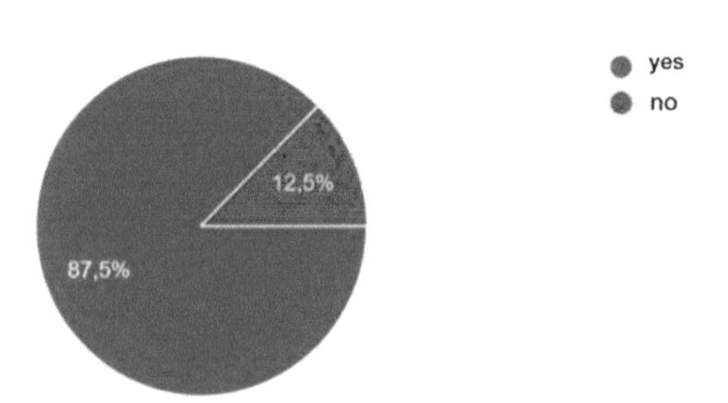

Fig. 4. Perceive of Flexibility

Table 8. Perceive Easy to Use.

Answer	Total Respondent Number	Percentage
Very Easy	11	35,5%
Easy	18	58,1%
Difficult	1	3,2%
Very Difficult	1	3,2%
Very Easy	11	35,5%

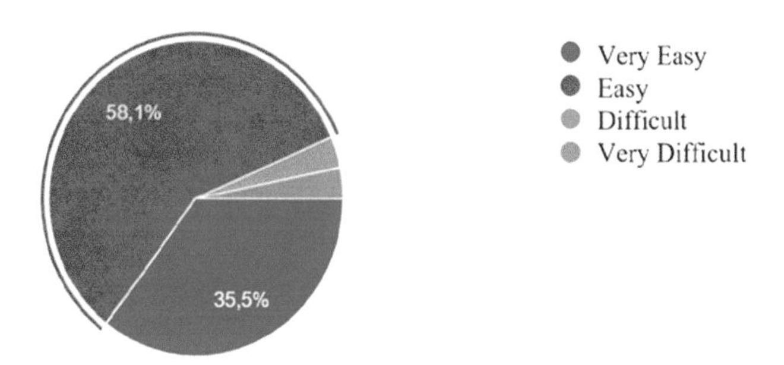

Fig. 5. Perceive Easy to Use

Table 9. Preferred Feature.

Answer	Total Respondent Number	Percentage
Chat	6	18,8%
Reaction	2	6,3%
Share screen	14	43,8%
Virtual Background	8	25%
None	1	3,1%
StickerEffect	1	3,1%

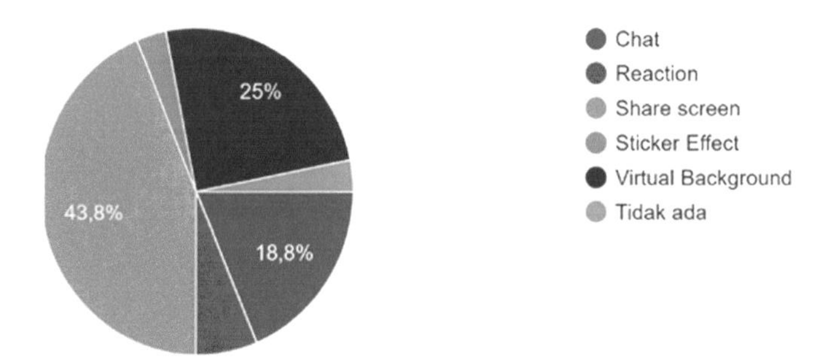

Fig. 6. Preferred Feature

4 Result and Discussion

During the epidemic, 32 respondents from different universities used the Zoom Meeting application to study online for the first time. However, our study indicates that using Zoom Meeting is successful, adaptable, and simple to comprehend for students engaged in online learning.

As shown in Table 1, the majority of respondents were in their third semester, with a total of 22 students. There were 4 students in Semester 1, 3 students in Semester 5 and 1 student in Semester 6 respectively.

According to the questionnaire given to various students from 15 different universities, the majority respondents (15 out of 31) were from Bina Nusantara University. The second-highest number came from Padjadjaran University and Brawijaya University, both with two as shown in Table 2.

Zoom Meeting is the most frequently used application by university students to conduct online learning as shown in Table 3 and Fig. 1. It is counted 25 respondents or approximately 78.1%. Following Zoom Meeting, there were those who use Google Meet, for 5 respondents (15.6%), and 2 respondents or 6.3% were using Microsoft Teams.

As shown in Figure Table 4, more respondents utilize the online learning software due to University Recommendation. There are a large number of students who utilize the Zoom Meeting media to conduct online learning as a result of the University's suggestion. However, some individuals use the Meeting Zoom program due to its ease of use as an online learning tools. As a result, we know that there is a necessity to have online learning tools, particularly Zoom, in order to effectively monitor learning during a pandemic situations.

It was discovered that 50% of respondents utilized video conferencing apps on a daily basis. Additionally, 43.8% of respondents utilize video conferencing three to four days per week, while 6.3% use video conferencing 1–2 days per week as shown in Table 5 and Fig. 2.

According these finding, half of our respondents use the Zoom Meeting Application. During a pandemic, when we are required to do everything from home, this Zoom Meeting Application is critical for connecting each of these fellow students whether for online lessons, group work, or for a meeting.

Table 6 and Fig. 3 shows that 56.3% of students felt that using the conference application was more beneficial for doing online learning, while only 43.8% disagreed. What makes Zoom Meeting less successful for distant learning are due to (1) regular issues such as weak of internet signal for students who do not utilize Wi-Fi (2) Difficulty with the practical lesson characteristic; (3) Frequently appearance of odd sound disturbances that interfere the learning activities. However, students find that utilizing the Zoom Meeting tool is very convenient. This is because lecturers and students communicate more effectively than paper based communication. Oral communication may be more easily understood than written communication. Additionally, the usage of the Zoom Meeting application may assist students in saving time and money.

Table 7 and Fig. 4 shows the perceive of flexibility of Zoom Meeting. Twenty-eight respondents said that the Zoom program was very flexible or versatile, while four respondents or 12.5% felt that it was less flexible. The Zoom Meeting is considered to be flexible because it can use anywhere and at any time, thus it is not time or location dependent. Existing characteristics such as virtual backdrop enable it to be more adaptable according to user's need.

According to Table 8, 35.5% of respondents said that Zoom Meeting was very easy to grasp, while 58.1% felt it was simple to comprehend. This indicates that nearly all respondents thought the Zoom Meeting Application was simple to use. This is possibly because Zoom Meeting has a straightforward design that prevents the user from being confused while using it. The concise delivery or explanation of features makes this program more understandable to many students who are now engaged in online learning.

With a total of 14 responses and a percentage of 43.8%, respondents selected the sharing screen as their favorite feature because it is effective and quick at presenting what Zoom users want, particularly during learning, presentations, and absences as presented in Table 9 and Fig. 6.

The share screen is one of the most helpful features; since a speaker or audience often encounters audio or sound interference, messages are not delivered correctly. The share screen enables a speaker to provide messages to the audience as it is. Thus, an audience that is unable to hear the speaker well may nevertheless view the speaker's screen display. This, however, needs collaboration between the speaker and the listener in order to deliver the message effectively.

Following the share screen function, the virtual backdrop feature had a 25% response rate, or 8 respondents. This feature enables users to customize the video dis-play's backdrop during a video call. This function is popular with many college students since it allows them to replace their messy living room with a beautiful landscape or any picture they want as a virtual backdrop.

The third popular element is Chat, which received six responses or 18.8% response rate. Several respondents indicated that they chose the chat feature because it allows them to communicate privately with other friends without being seen by the host or other people other than the intended person. Chat can also help facilitate communication if the ohmic one has too much voice, as the chat feature can assist.

Reaction feature enables us to ascertain the expression of each person who participated in the learning. Additionally, by using this functionality, we may provide an entirely unique answer. Users are not required to express theirs feelings to the other person. Simply send a clap or thumbs up emoji by choosing the 'Reactions' button at the bottom of the page and selecting the appropriate emoji. The emoji will display on the screen for 5 s before disappearing. There were two respondents that selected the reaction feature or 6.3%.

Finally, there is the sticker effect function. By using augmented reality (AR) technology, this feature enables users to apply a variety of fascinating and humorous face filters, enhancing the fluidity and excitement of virtual meetings. Zoom's numerous video filters are arguably identical to the augmented reality face filters found on major social networking platforms such as Line, Snapchat, and Instagram. Our responders like this feature since it is distinctive and it helps to break up the monotony of studying.

5 Conclusion and Recommendation

During the COVID19 pandemic, learning has grown more difficult since face-to-face instruction is not permitted at many institutions. However, there is now a way to ensure that the learning process goes well, and that is via the use of video conferencing tools, one of which is ZOOM Meetings.

We found that all respondents utilize the Video Conference application as a medium of instruction. Zoom meeting is the most well-known video conference application utilized during the online learning activities. The majority of them said that the application enhanced the teaching and learning process. However, several students responded that using the video conferencing program was ineffective due to the flaws and the fact that many parts of their houses had inconsistent internet signals.

Students utilize the Video Conference program for a variety of reasons, one of which is because their institution requires them to do so as an online learning medium. The majority of them use video conferencing apps daily as a mode of instruction. Another reason is that the program is very simple to comprehend.

Additionally, some features are an incentive to use the Video Conference application. The Share Screen option is one of the most popular aspects among the users. Video Conferencing program offers a number of benefits and drawbacks. The drawbacks that are often faced by students is being very wasteful with data limits, which is rather expensive for students who do not utilize a Wi-Fi. ZOOM Meetings was also reported to have had a large amount of data exposed caused the consumers disappointment. There is something that the majority of students dislike the ZOOM Meetings program, it is mainly because it has a time restriction, which often impedes student learning.

Behind those limitations, ZOOM Meetings offers several benefits, including a big enough room capacity for students to host their own events. ZOOM Meeting has a plethora of intuitive features, which serves as one of the application's primary draws. Additionally, these apps are available on a variety of devices, making them more accessible to students who do not own computers. This application is very beneficial for those who want to conduct a presentation.

References

1. Serhan, D.: Transitioning from face-to-face to remote learning: students' attitudes and perceptions of using Zoom during COVID-19 pandemic. Int. J. Technol. Educ. Sci. **4**(4), 335–342 (2020)

2. Asfar, A.M.I.T., Asfar, A.M.I.A.: The effectiveness of distance learning through Edmodo and Video conferencing Jitsi meet. J. Phys. Conf. Ser. **1760**(1), 012040 (2021). https://doi.org/10.1088/1742-6596/1760/1/012040
3. Bozkurt, A., Sharma, R.C.: Emergency remote teaching in a time of global crisis due to CoronaVirus pandemic. Asian J. Dist. Educ. **15**(1), i–vi (2020)
4. Mishra, L., Gupta, T., Shree, A.: Online teaching-learning in higher education during lockdown period of COVID-19 pandemic. Int. J. Educ. Res. Open **1**, 100012 (2020)
5. Video Conferencing, Cloud Phone, Webinars, Chat, Virtual Events | Zoom. https://zoom.us/. Accessed 15 Dec 2021
6. Alnemary, F.M., Wallace, M., Symon, J.B.G., Barry, L.M.: Using international videoconferencing to provide staff training on functional behavioral assessment. Behav. Interv. **30**(1), 73–86 (2015)
7. Scanga, L.H., Deen, M.K.Y., Smith, S.R., Wright, K.: Zoom around the world: using videoconferencing technology for international trainings. J. Ext. **56**(5), 14 (2018)
8. Haqien, D., Rahman, A.A.: Pemanfaatan zoom meeting untuk proses pembelajaran pada masa pandemi COVID-19. SAP (Susunan Artik. Pendidikan) **5**(1) (2020)
9. Nguyen, T.: The effectiveness of online learning: Beyond no significant difference and future horizons. MERLOT J. Online Learn. Teach. **11**(2), 309–319 (2015)
10. Jolliffe, A., Ritter, J., Stevens, D.: The Online Learning Handbook: Developing and Using Web-Based Learning. Routledge (2012)
11. Huang, H.: Instructional technologies facilitating online courses. Educ. Technol. 41–46 (2000)
12. Anderson, T.: The Theory and Practice of Online Learning, 2nd edn. Athabasca University Press (2008)
13. Archibald, M.M., Ambagtsheer, R.C., Casey, M.G., Lawless, M.: Using zoom videoconferencing for qualitative data collection: perceptions and experiences of researchers and participants. Int. J. Qual. Methods **18**, 1609406919874596 (2019)
14. Gray, L.M., Wong-Wylie, G., Rempel, G.R., Cook, K.: Expanding qualitative research interviewing strategies: Zoom video communications. Qual. Rep. **25**(5), 1292–1301 (2020). https://doi.org/10.46743/2160-3715/2020.4212
15. Dumford, A.D., Miller, A.L.: Online learning in higher education: exploring advantages and disadvantages for engagement. J. Comput. High. Educ. **30**(3), 452–465 (2018). https://doi.org/10.1007/s12528-018-9179-z

Proposal for a Theatre Optique Simulated Experience Application

Nanami Kuwahara(✉) and Takayuki Fujimoto

Toyo University, Kujirai 2100, Kawagoe, Japan
s3B102100038@toyo.jp

Abstract. Charles Emile Reynaud's invention, Theatre Optique, is essential to the history of pre-cinema. Theatre Optique is a device to project animations, and Reynaud used it to show narrative films. Moreover, Reynaud not only developed the equipment, but also created the story, drew the pictures, and held the performances. Today, videos can be easily planned and edited with Premiere Pro and other software, but it is difficult to get the sense of creativity that comes with the limited materials of film. Reynaud was able to check the reaction of the public while he screened his films at the Theatre Optique. For this reason, the technicians operating the equipment were required to have the ability to "tell a story" (directing ability). These skills and senses are being lost due to the decline of the equipment and the development of technology. Therefore, we will re-construct the value and significance of the Theatre Optique in pre-cinema history and culture. Furthermore, we will design an application that can simulate the experience of the Theatre Optique.

Keywords: Theatre Optique · Simulation Experience System · Visual Media Expression · PreCinema

1 Background

Nowadays, people are bombarded with images and video content. There are many types of production and methods, such as documentary video, animated movie, and 3DCG. Animated films in particular have had a great impact on the current Japanese "Anime" culture. The "Invention of Cinema: Pre-Cinema" is an important part of the history and origin of animated films. Pre-cinema is a general term for the history and culture before the birth of cinema. The birth of cinema refers to the origin of the cinematograph, invented by the Lumière brothers. In this pre-cinema era, various visual expression devices were developed in the invention of cinema. For example, the zoetrope, kinetoscope, and the Theatre Optique were invented. In particular, the Theatre Optique was the last moving image device in the pre-cinema era. For this reason, the Theatre Optique plays an important role when discussing Pre-Cinema. Moreover, it can be said that the Theatre Optique is a core device for tracing back the history of moving images such as film and animation.

T. Matsuo et al. (Eds.): AIMD 2019, LNNS 677, pp. 48–58, 2023.
https://doi.org/10.1007/978-3-031-30769-0_5

The Theatre Optique is a movie device invented by Emile Reynaud in 1888. The Theatre Optique combines a moving image of the characters with a background film to play animations. In addition, the reels are turned to create a theatrical effect. The development of the Theatre Optique was based on Reynaud's invention, the Praxinoscope. In 1877, the Praxinoscope was invented as a moving image device for person-al viewing. This 'personal' device was transformed into a 'mass' device. In addition, the "repeated" moving image was transformed into a "narrative" moving image. Thus, the Praxinoscope was evolved into the Theatre Optique, which became the forerunner of animated movies.

Reynaud used the Theatre Optique to show short films of about 600–700 frames for about 15 min. Reynaud adjusted the limited length of the film using only a handle. During the screening, he watched the audience's reaction and developed the story. This shows that he made the most of the short film to share an entertainment experience. However, with the release of the cinematograph invented by the Lumière brothers in 1895, the Theatre Optique lost its audience. The Theatre Optique did not become widespread because of the complicated operation of the equipment and its high price. Furthermore, there are no existing devices of the Theatre Optique. For this and many other reasons, the Theatre Optique is not well known to the world today.

As shown, the Theatre Optique contributed greatly to the development of animated movies and popular entertainment. Nowadays, people can see how the Theatre Optique works and watch films played on this device on video playback sites. However, there are very few imitations of this device, so people cannot experience it for themselves. Therefore, we will design an application system to reproduce the Theatre Optique by miniaturizing this large and expensive device for more easy and casual experience (Fig. 1).

Fig. 1. Theatre Optique "Pantomimes Lumineuses" (1892).

2 Purpose

In this study, we will design an application that imitates the "Theatre Optique", a moving image device developed by Emile Reynaud in 1888. We will design an application that allows users to easily simulate the past medium, Theatre Optique.

By incorporating these past technologies into the application, it can be used as educational content to learn and review the history of media. The devices and toys born in Pre-Cinema had a great impact on people's visual expression. Therefore, it is important to understand how the technologies of the past were developed and used in media education.

In the system proposed in this study, short films ranging from a few frames to 700 frames are played back manually while narrated like a picture-story show. Unlike modern playback systems that play the film with the push of a button, the Theatre Optique system requires sounds in real time and demonstrates the work on the spot. This can be said to provide a unique work and experience by expressing everything at the same time. Therefore, it is possible to provide PreCinema experience by re-creating the technology used in the past, such as the Theatre Optique.

3 Relevant Study

In this chapter, we present examples related to the Theatre Optique and the proposed application. Unlike live-action video images, the Theatre Optique is an animation device that moves pictures. People tend to feel nostalgic when they come in contact with things and experiences that they used in the past. Therefore, various products and content have been created to provide 'nostalgia'. In many cases, such products incorporate evolved technologies to recreate the technologies of the past. From this point of view, we will examine related cases.

3.1 Flip Book

A flip book is a cartoon comic, a comic book in which pictures are drawn on multiple layers of paper and the picture appears to move when the paper is quickly flipped. Since the pictures appear to be moving, it is related to other visual techniques such as moving images and animation. The inventor of the Theatre Optique, Reynaud, drew directly on film and moved the characters frame by frame. Although the materials used in flip books and the Theatre Optique are different, they are the same in that they work within analog limitations. With digital products, it is easy to duplicate a single frame, but with analog products, each frame must be drawn from scratch. This requires a lot of time and effort, which sometimes prevents the production of many works. Moreover, as in the case of the Theatre Optique, it is possible to adjust the speed at which the paper is turned to create a gradual progression of the story. While there are many similarities, the target audience for paper-based flip books is from one to a few people. It is not possible to show animations to a huge audience at once like the Theatre Optique.

3.2 The Application of Animation Creation Using iPad

"Looom" by iorama is an application for creating hand drawn animations. Being influenced by music production tools, it allows users to create animation as intuitively as playing a musical instrument. In a flip book, people use paper and pen to create analog moving pictures. Looom, on the other hand, is a tool for creation on a digital device. Looom is also unique in how the application operates. Users draw pictures with their right hand using their fingers or a pen, while their left hand spins a reel that adjusts the animation frame. Spinning the reel as the animation progresses is not only intuitive, but also reminiscent of films used for the devices such as the Theatre Optique. In this way, the birth of digital terminals has made it possible to create animation with ease.

3.3 Experience Applications for Simulating the Analog Devices

Analog devices and technologies used in the past are occasionally reproduced in applications. The analog devices of the past include products such as rotary telephones, vinyl records, and film cameras. With the spread of computers and the advancement of technology, the number of users of these products has decreased due to their inferior convenience and functionality. However, there are a number of people today who continue to use the technologies and products of the past. They continue to use these products for a variety of reasons, such as feeling nostalgic or experiencing the "goodness" of analog. Many applications reproduce the analog technologies of the past because of this demand.

For example, there is a music playback application that reproduces a record player, such as "Retro Player" (STUDIO-KURA). People can adjust sound distortion, skipping, noise, which are the characteristics of vinyl records, and listen to music in the style of vinyl records. There are other applications that reproduce the experience of cassette tapes, such as "Magnetola". While there are other music playing applications designed to look like cassette tapes, Magnetola has been developed to reproduce the smallest details, from the switches to the reels. For example, the amount of tape changes depending on the length of the song, and when the playback speed is in-creased, the tape can be heard winding rapidly. Not only does it look like a cassette tape, but it also performs like a player.

These applications not only provide a visual representation of the old equipment, but also give the user a sense of what it was like to use such equipment. Users who have used vinyl records or cassette tapes will find them nostalgic. In addition, it can be used as a pseudo-experience application for users who have never used or know nothing about them. It is thought that the ease of experiencing them on a smartphone will lead to a better understanding of the media than using the actual product.

3.4 CRT-Like LED TV

The 'VT203-BR' is a product that reproduces CRT televisions that were widely used in the past. It is a reproduction of the cathode-ray tube, which is not in vogue now, in the form of an LED LCD TV. The design of the product is based on the cathode-ray tube TVs that appeared in the 1970s. While 3.3 was an example of an application reproducing technologies used in the past, the VT203-BR differs from these in its mechanism.

Although VT203-BR looks like an 'old' or 'retro' TV, its performance is the same as that of the latest TVs. The VT203-BR is a "new" TV in the sense that it has realized a retro television that was unlikely to exist. In addition, details such as channel dials and switching switches have been reproduced, making it intuitive to operate. The VT203-BR is a product that gives the user a sense of nostalgia not only from the feel of operating a TV of the past, but also from its design. In this way, the latest technology replaces devices from previous generations, creating new value by expressing their designs and details.

3.5 Significance of the Proposed Application

Based on the above prior work, it can be said that the proposed application has significance to reproduce and experience the Theatre Optique. Previous imitations of the Theatre Optique have been produced, but the period and location of the experience have been limited, and the history and value of this device could not be easily experienced. However, just as there are applications that mimic vinyl records and cassette tapes, we believe that the Theatre Optique can be made into an application that al-lows anyone to experience the technology of the past.

Recreating and experiencing the last devices in pre-cinema with this application will lead to an understanding of media history. In addition, there is no application that expresses the Theatre Optique. Therefore, we can say that this study is highly novel.

4 Overview of the Application

This application recreates the pre-cinema Theatre Optique. In other words, it is an application for experiencing the Theatre Optique. Therefore, the functions of the application are quite simple. The features of the application are as follows:

-Rotate the film on the screen to advance the story.

- The film has two layers, the front and the background.
- Projection method is wired (cable) or wireless (Wi-Fi) connection to a display.
- This application cannot incorporate audio.

The details of the mechanism and functions of the Theatre Optique simulated experience application are described below.

5 Application Mechanism

In this chapter, we explain the mechanism and functions of the proposed application. In this study, we will use a digital terminal to play a movie utilizing the 'Theatre Optique'. In order to reproduce the Theatre Optique in the application, the following flow will be realized.

(1) Select a character film (A).
(2) Select the background film (B).
(3) Combine (A) and (B).
(4) Turn the reel to play the animated movie.

Figure 2 shows the transition diagram of the application screen. The total number of screens in this application is four.

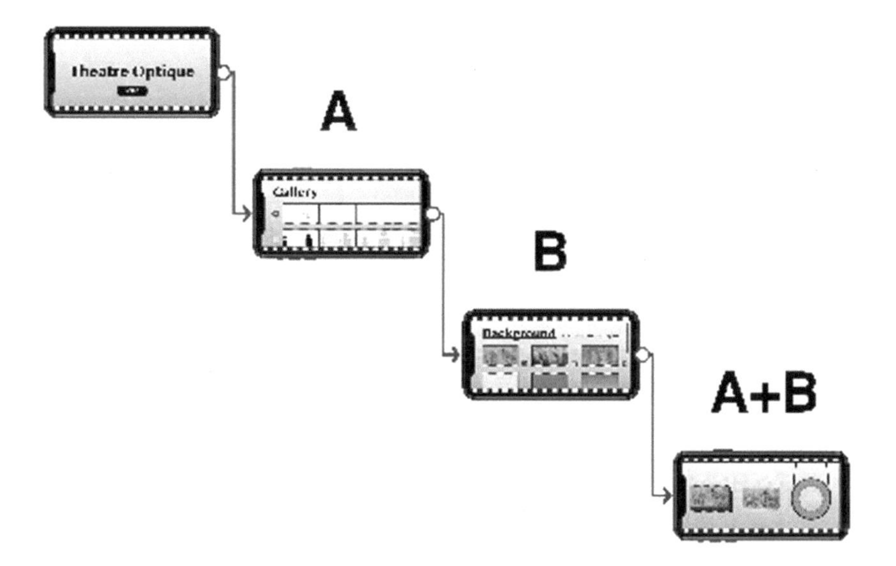

Fig. 2. Transition Diagram of the Application Screen.

5.1 Initial Screen of the Application

Figure 3 shows the initial screen of this application. By tapping the "Start" button, the user will move to the next screen (5.2).

Fig. 3. Initial Screen of the Application.

5.2 Character Film Selection Screen (Gallery)

Figure 4 shows the screen for selecting materials for 'Storytelling'. The gallery display is designed to continue downwards. By scrolling down, users can see and select other films. In the Theatre Optique, only one film to move characters can be installed. Therefore, the character film.

Selection screen in this application also allows users to select only one film. Tap the radio button to select a film. The character film will be at the front of the movie when it is played back (Fig. 4).

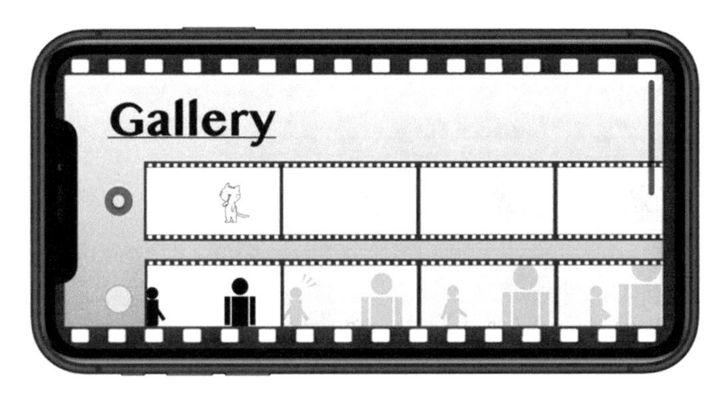

Fig. 4. Image of the Screen for Selecting a Character Film.

5.3 Background Film Selection Screen (Background)

Figure 5 shows the selection of material for 'Telling a story'. By scrolling down, the users can check and select other background films. A list of backgrounds is displayed with a tick button, and multiple selections can be made from the display. Users can prepare the background for the movie to be played in the application. In Theatre Optique, the background film was installed separately from the reeling film. There-fore, it was possible to use more than one type of background film in Theatre Optique. Reflecting this, it is possible to select multiple background films in the application.

Fig. 5. Image of the Screen for Selecting a Character Film.

5.4 Animated Movie Playback (Projection) Screen

In the movie playback (projection) screen, the films selected in 5.2 and 5.3 are combined to play an animated movie (Fig. 6). When projecting at the Theatre Optique, the character film and the background film were set up separately for operation. This is a mechanism similar to current celluloid films. In other words, it is possible to switch the scenes of the story by replacing the background film. However, to strictly incorporate this feature into the application, one device for the character film and another device for the background film would be required. This is a far cry from the ease of the Theatre Optique experience. Therefore, we designed an operation screen that allows two types of film operation with a single digital terminal.

From left to right, the operation buttons are arranged as follows: background film button, demonstration movie screen, and character film reel button. Basically, operation is performed with both hands. The left hand operates the background screen, and the right hand plays the movie. The demonstration movie screen in the center shows the same as the projected movie in real time.

The background film button allows users to change the background by tapping or swiping the screen. These backgrounds are the multiple backgrounds selected in 5.3 and can be used to switch scenes to develop the story.

In the demonstration movie screen, the moving image that is being played (projected) on the display is shown in real time. By playing back the movie while watching this screen, users can check at hand how the background film and the character film are projected. For example, even if the user cannot directly see the projection display during playback, the user can check the demonstration screen and proceed with the story.

With the character film reel button, turning the film (reel) button with the user's finger will play the movie. Turning this button counterclockwise (leftward) plays the film and turning it clockwise (rightward) rewinds the film. The Theatre Optique had two large disks, and the right hand fed the film and the left hand rewound it. This was the shape of the Theatre Optique because it was a large-scale device. However, the application is not designed to express the size of the device. The purpose of this study is to reproduce the mechanism. Therefore, in the application, the character film reel but-ton can be used to feed/rewind the film. Furthermore, the speed at which the reel button plays the movie changes depending on the speed at which the reel is turned. Turning it quickly allows users to move the characters in an instant, while turning it slowly slows down the progress of the story. Turning the reel clockwise (clockwise) rewinds the film, allowing users to play the same scene repeatedly. At first glance, it looks like a manual animated movie playback application, but it is a system that allows users to demonstrate their expressive power and creativity to entertain audience depending on how they turn the reel.

5.5 How to Connect the Mobile Device to the Display

This application realizes a media device for the masses as well as the Theatre Optique. There is a risk that smartphones and tablets will become movie playback devices only for individuals or small groups. Therefore, they need to be connected to displays such as TVs, desktop PCs or projector screens. In this study, we propose two connection methods.

There are two ways to project this application (to connect to the display for movie play): wired and wireless (Fig. 7). The first is to connect the mobile device to the display with a cable (wired). The second method is to connect the mobile device to the display via Bluetooth or Wi- Fi (wireless). These are different from the methods of controlling the display with a tablet, such as remote desktop or screen mirroring. Instead, these are similar to a connection method using a presenter's tool such as cases with PowerPoint. Connecting the terminal to the display in this way makes the display a sub-display. Another example is the connection between an LCD tablet (drawing tablet) and a PC, where the screen is extended, and the movie is played in a way that switches the mapping screen.

Fig. 6. Image of the Movie Playback (Projection) Screen.

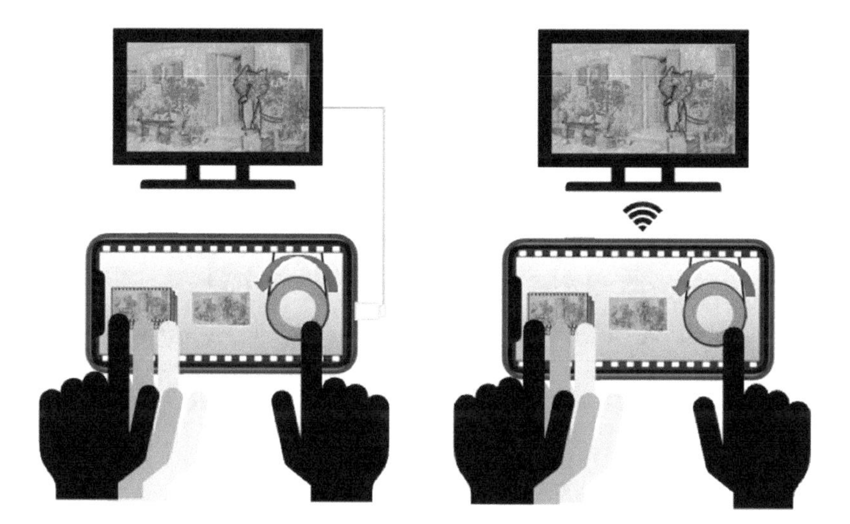

Fig. 7. Image of a Mobile Device Connected to a Display.

6 Conclusion and Future Works

In this paper, we proposed a design for a simulated experience application that replicates the Theatre Optique. Until now, the Theatre Optique was a moving image de-vice from the late 19th century. Moreover, the original devices do not exist and few imitations exist. However, the Theatre Optique is an important device in media history, as it is the starting point in animation movies and mass entertainment. This led to our proposal for an application to recreate the lost culture and technology of the Theatre Optique.

In this study, we only designed a proposal for the application, however, we will implement the application in the future. We will also conduct evaluation experiments with subjects to examine the practicality of the application. In the future, we would also like to examine whether it is possible or necessary to include the function to create a story from the very beginning in this application. This consideration is due to the fact that the inventor, Reynaud, drew pictures on the film by hand from scratch. A reliving system that recreates the mechanisms of past devices and even requires users to produce a story from scratch, may have plenty of meaning for users.

This application can be a possible breakthrough to provide an opportunity to reconsider the history of pre-cinema devices and media.

References

1. Cholodenko, A.: The Animation of Cinema, The Semiotic Review of Books, vol. 18.2 (2008)
2. Morioka, Y.: Between Magic and the Algorithmic Image Zentrum fur Kunst und Medientechnologie, pp.45–50 (1994)
3. Ceram, C.W.: Eine Archaologie des Kinos. Rowohlt, Verlag, Hamburg (1965)
4. Yoshioka, A.: Visual media in science education in japan viewed from learning theoretical and historical approach. Rikkyo University annual report of the Department of Education, No.58, pp. 111–139 (2015)
5. Simon, G., Ota, Y.: Simon ganahl 'from media archaeology to media genealogy: an interview with erkki huhtamo.' Cult. Sci. Rep. Kagoshima Univ. **85**, 15–23 (2018)
6. Australian Centre for the Moving Image, "Théâtre Optique", Australian Centre for the Moving Image. https://www.acmi.net.au/works/100579--theatre-optique/. (Referenced 2022–01- 17)
7. G. Sadoul, "Histoire générale du cinema1, L'invention du cinéma 1832–1897", Denoël, 1973, 574p.
8. iorama, "Looom". https://www.iorama.studio. (Referenced 2022–01–20)
9. STUDIO-KURA, "Retro Player". http://www.studio-kura.com/index.html. (Referenced 2022–01–20)
10. Hovik Melikyan, "Magnetola". https://hmelik.wixsite.com/melikyan/magnetola. (Referenced 2022–01–20)

A Classification and Analysis Focusing on Attempts to Give a Computer a Personality: A Technological History of Chatbots as Simple Artificial Intelligence

Taishi Nemoto[1,2](✉) and Takayuki Fujimoto[1]

[1] Toyo University, 2100 Kujirai, Kawagoe, Japan
nemotouc@gmail.com
[2] Kagoshima Women's Junior College, Komacho 6-9, Kagoshima, Japan

Abstract. In recent years, research on artificial intelligence has been thriving. The use of AI by general consumers is rapidly expanding. Among them, the most popular AI would be chatbots. Chatbot systems, which have been proposed since the 1980s based on the image of general-purpose artificial intelligence, have simple mechanisms, but many of them have a high degree of human-likeness. Voice assistants installed in smartphones are often based on chatbot technology, and the digital native generation casually uses them on a daily basis. Chatbots are used in a variety of fields besides business and education and are expected to be used more and more in the future. The number of such chatbots is increasing all over the world, and the number of services is increasing rapidly. The number of users and characteristics of these services are changing day by day, and Mega-Techs are building chatbot platforms one after another within a short period of time. The market is growing steadily. The needs and applications of chatbots are quite diverse, for example, they do not only answer simple questions but also recommend products or guide payments. In this paper, we will overview history of simple artificial intelligence systems, from ELIZA, the ancestor of chatbots, to the present day's, and classify them in terms of mechanism and structure. We will also conduct a survey on people's attitudes toward AI and chatbots, in order to highlight the challenges with the future system of chatbots. In fact, many people argue that chatbots are merely displaying search results from a web browser. In this research, we organize and analyze the existing chatbots of various scales and their services.

Keywords: AI · Chatbot · Social media

1 Introduction

"Chatbot" is a coined term made of "Chat" and "Bot". It refers to a system that automatically converses with the user through text or voice. The term "chat", refers to re-al-time communication over the Internet, which is a two-way exchange of text. Messenger, bulletin board sites, and slack are the most famous services. "Bot" is an abbreviated word

T. Matsuo et al. (Eds.): AIMD 2019, LNNS 677, pp. 59–70, 2023.
https://doi.org/10.1007/978-3-031-30769-0_6

for "Robot", a program that automates tasks and processes on behalf of humans. Google uses bots, called crawlers, to go from link to link on web pages to determine if the applied information is useful or not. It is said that these chatbots can improve operational efficiency of the call centers and other handlings that require operators. However, they are snot capable of responding perfectly like a human concierge, yet.

At present, the chatbots' use ranges from messenger bots to the use as customer support for products. Today, they are mostly used in voice assistants for smartphones. In terms of technology and mechanism, the voice assistants are almost the same as chatbots. Strictly speaking, there are some voice assistants that are different from AI, but it can be said that the voice assistants is the AI service that has the largest number of users at present. The content of the service is mainly brief Q&A, and its simplicity of the service has been well attuned by the users. Basically, there are two types of mechanisms: rule-based type (artificial incompetence) and machine learning (artificial intelligence). Each has its own merits and demerits.

1.1 Rule-Based Chatbots

The mechanism of a rule-type (artificially incompetence) chatbot is simple, and it has many kinds of limitations. This type of chatbot called 'scenario type', automatically responds to pre-created scenarios and carries on conversations. The simplest type of chatbot, called a 'dictionary chatbot' also falls into this category, and it returns a specific sentence when an exact keyword is entered. Since they are programmed to answer according to a scenario, the system scarcely answer incorrectly. On the other hand, this point also can be the disadvantage as limited capability of answering only according to the rules, and the cost tends to become high because to have it answer many questions, setting up a huge number of rules is necessary.

1.2 Machine Learning Chatbots

A machine-learning chatbot displays more appropriate answers based on accumulated data that has been learned in advance. They can cope with fluctuations in expressions (subtle difference in wording) that vary from the questioner to questioner. The more frequently users use them, the more response data they can collect and learn from, and it leads to increase the system accuracy. In some cases, a certain length of learning period is required to improve the accuracy and sometimes, unexpected answers may be given to the users. Even if the model has already been trained, it cannot be completely autonomous, and it needs to be tuned by human intervention to some extent.

2 Background

Currently, there are a lot of chatbot-related researches, as well as the products and ser-vices created by companies. Not only are they used for business purposes such as customer support, Q&A, and product recommendation systems, but also for a variety of scenes such as medical response including COVID-19 treatment, disaster relief, education, and communication. This is because chat technology, coupled with machine learning in the third AI boom, is advancing constantly. Therefore, we thought it is necessary to summarize not only the world-famous AI cases but also the chatbots provided by various IT companies. The market size of chatbots has been rising steadily. According to a report by MARKETS AND MARKETS, the interactive AI market is expected to grow at a CAGR of 21.9% from US$4.8 billion in 2020 to US$13.9 billion in 2025. This is due to the COVID-19 pandemic and the DX trends. It is be-coming more and more common that investment for AI and its infrastructure is more financially efficient than for human resources. Upon such tendency, more and more companies are adopting 'chatbot's as a service. In the first place, chatbot services have emerged in the last decade and have penetrated rapidly. In particular, around 2016, mega-tech companies such as Microsoft and Meta (formerly Facebook) start-ed providing platforms, and it functioned as a direct trigger. The conversational API (Application Programming Interface) allows the implementation of user-customizable autonomous bots. The following is a list of major companies and their services related to chatbots (Table 1).

Table 1. Companies and their chatbot services.

Corporation	Service
Microsoft	Skype Bot Platform (2016)
IBM	IBM MobileFirst for iOS (2016)
Meta (ex: Facebook)	Facebook Messenger Platform (2016)
slack	Slack API (2016)
Google	Allo (2016–2019)
LINE	Messaging API (2016)
Twitter	Twitter API (2009)

3 Purpose

The most common type of AI today is the chatbot. This is because an increasing number of companies are adopting chatbots. With the rapid rise in the market size and services over the past few years, gradually various services at different level of scales also appear to be growing. The continuous increase is expected in the next decade. Under current circumstances, the purpose and usefulness of these services covers a log of ground, and they also vary greatly in quality. Therefore, I would like to examine whether chatbots are really useful in daily life and business by looking at the history of existing chatbots. Thus, the purpose of this research is to organize various chatbots and services, both large and small scales, and analyze the types of them in terms of mechanisms and characteristics. Although chatbots and voice assistants imitate humans, they do not perfectly reproduce human responses. This is because their reactions are only based on output from a database (dictionary). Even if the reply words are changed by pattern recognition using machine learning, it does not mean that the system truly understands human language. On the other hand, there is also a phenomenon called the 'ELIZA effect', upon which the users assume that computer behavior resembles human behavior. By overviewing the history of such chatbots, we classify and analyze it specifically focusing on the attempts to reproduce human likeness on machines.

4 Representative Services

In this chapter, we summarize the major chatbots to the date. We will cover the major systems, from globally prevailing products to the original of every kind of chatbot. In addition, voice assistants, which are technologically related to chatbots, will be discussed and summarized based on characteristics. There are some people who think of messenger bots or small windows on a website for customer support as "chatbots", but voice assistants can be also classified as chatbots.

4.1 ELIZA

ELIZA, developed by Joseph Weizenbaum in 1966, is considered to be the ancestor of chatbots. ELIZA is a natural language processing program that uses a simple pattern matching technique. For a single keyword, multiple conversation patterns are stored in a database, and the program outputs them in order. The following is an image of ELIZA (Fig. 1).

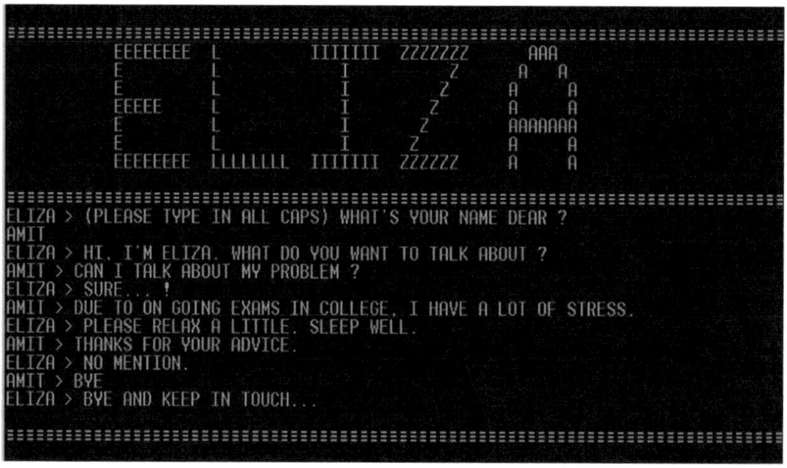

```
ELIZA > (PLEASE TYPE IN ALL CAPS) WHAT'S YOUR NAME DEAR ?
AMIT
ELIZA > HI, I'M ELIZA. WHAT DO YOU WANT TO TALK ABOUT ?
AMIT > CAN I TALK ABOUT MY PROBLEM ?
ELIZA > SURE...!
AMIT > DUE TO ON GOING EXAMS IN COLLEGE, I HAVE A LOT OF STRESS.
ELIZA > PLEASE RELAX A LITTLE. SLEEP WELL.
AMIT > THANKS FOR YOUR ADVICE.
ELIZA > NO MENTION.
AMIT > BYE
ELIZA > BYE AND KEEP IN TOUCH...
```

Fig. 1. ELIZA chat screen.

Originally, ELIZA was developed focusing on the treatment for patients with mental illness that can make them recover by themselves proactively, by paying attention to their worries and feelings of anger. ELIZA's function of listening to what the person was saying, asking core questions, and guiding the person not to stray from facing their worries played the role of a psychiatrist or counselor. It is said that some of the patients who were exposed to ELIZA became emotionally involved, and some even confided their problems to the program. This interaction gave rise to the term "ELIZA effect," in which people unconsciously assume that the computer behavior is similar to a human behavior, even though they know the object is a computer. This ELIZA was built up in various environments and is thought to have been the basis of subsequent AI research such as A.L.I.C.E. and Siri.

4.2 PARRY

In 1972, Kennel Colby, an American psychiatrist, published a system that modeled the behavior of schizophrenic patients and it was programmed to play a specific person. Like ELIZA, PARRY searched for keywords in a database and automatically responded to them. PARRY and ELIZA were different in terms of that PARRY was programmed to actively draw others into conversation by mentioning PARRY's own beliefs, fears, and concerns. In 1972, PARRY tried to interact with the existing ELIZA, but the interaction did not go well, and the conversation never made sense. PARRY's program was a bit more advanced than ELIZA 's, which simply asks questions about what the user has typed, but it turned out that PARRY could only output the expectable answer to a given question, without really understanding the meaning of what the conversation partner is saying. The following is an image of PARRY at-tempting to have a conversation with ELIZA (Fig. 2).

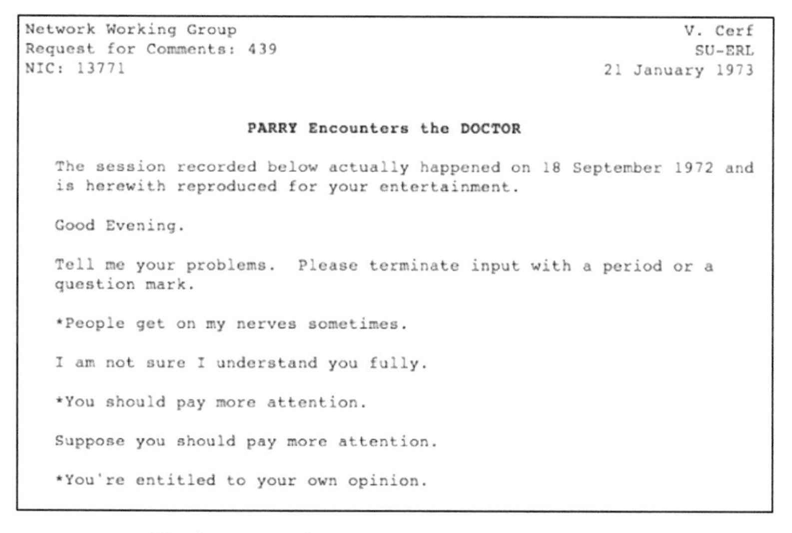
Network Working Group V. Cerf
Request for Comments: 439 SU-ERL
NIC: 13771 21 January 1973

PARRY Encounters the DOCTOR

The session recorded below actually happened on 18 September 1972 and is herewith reproduced for your entertainment.

Good Evening.

Tell me your problems. Please terminate input with a period or a question mark.

*People get on my nerves sometimes.

I am not sure I understand you fully.

*You should pay more attention.

Suppose you should pay more attention.

*You're entitled to your own opinion.

Fig. 2. Image of PARRY encountering with ELIZA.

4.3 Jabberwachy

This is a talking bot developed by Rollo Carpenter in 1988. The goal of this development project was to "simulate natural human conversation in an interesting, entertaining, and humorous way," and the objective was to pass the Turing test. It was an artificial intelligence development effort with a focus on human interaction. This system differs from other artificial intelligence in that it learns language and context through interaction with humans. The system is designed to work solely through AI-user interaction. It stores bot comment data from past conversations with users and uses them to find the most appropriate responses. In other words, the system builds a large database of contextually appropriate conversations. If enough information is provided, it was considered to be capable of learning to use slang, tell jokes and play with words. The system has participated in Loebner Prize Contest every year since 2003, achieving excellent results. The programs George (2005) and Joan (2006), on which the Jabberwacky system was based, won the Bronze Award. The problem is that it does not hold natural conversations that can satisfy everyone, although the system was designed to respond appropriately. This is because humans can say something in the wrong context in conversations, or may not answer questions. These cases will be stored in the database as out-of-context conversations. In addition, frequent changes in the subject make it difficult for the system to respond to them. The chatbot systems have been available on the Internet since 1977 and are still widely used.

4.4 Dr Sbaito

It is a chatbot developed by Creative Labs in Singapore for MS-DOS in 1992. It is similar to ELIZA in that it simulates a psychologist, but its novelty lies in the fact that it speaks using speech synthesis. By using the attached sound card, it successfully read out the text entered following the word "say". However, it was difficult for the chatbot to carry out complex conversations, and if the user uses offensive language or verbally abuses the chatbot repeatedly, it would cause an error, resetting the system. Its synthesized speech was a precursor of today's voice assistants but was not practical at the time. The following is an image of Dr. Sbaitso (Fig. 3).

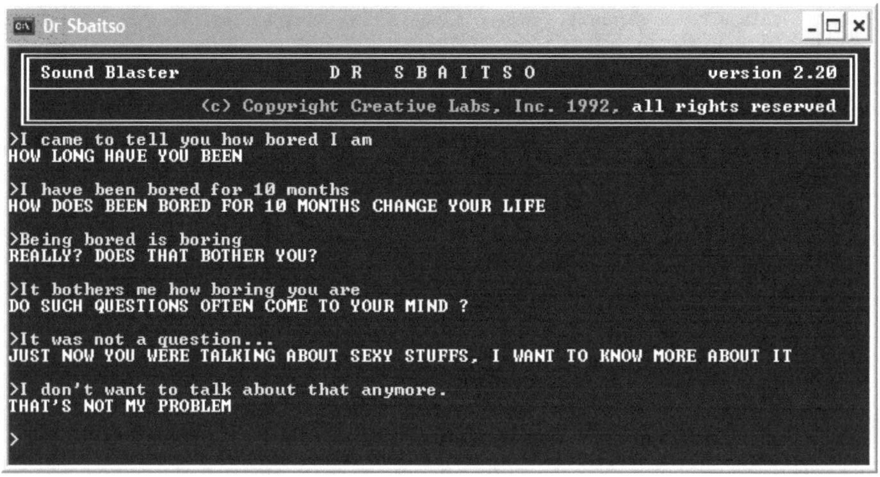

Fig. 3. Image of Dr. Sbaitso chat screen.

4.5 A.L.I.C.E

A.L.I.C.E. is a chatbot inspired by ELIZA. Richard Wallece started its development in 1995. A.L.I.C.E. is an acronym for Artificial Linguistic Internet Computer Entity. In 1998, it was rewritten in Java and later open-sourced. An innovative idea of using AIML (artificial intelligence markup language) was experimented. AIML is an XML schema which describes the conversational rules of artificial intelligence. The ALICE chatbot relies on predefined templates for interaction and has limited learning capabilities. Instead of using more advanced NLP techniques, it uses a set of rules to match the user's input against patterns to determine how to respond. However, the ALICE chatbot handles complex language well enough to generate very natural conversations. The AI chatbot has been updated repeatedly and is still in use. The following is an image of the ALICE chatbot (Fig. 4).

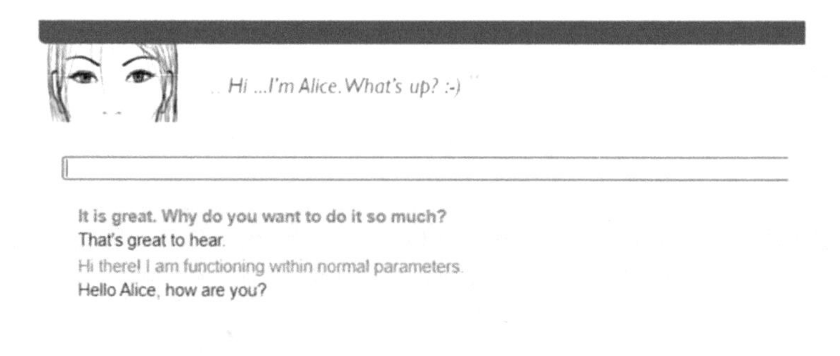

Fig. 4. Image of A.L.I.C.E. chat screen.

4.6 SmartChild

SmarterChild was developed by ActiveBuddy in 2001. Since the chatbot was available to use on AOL Instant Messenger and MSN Messenger, it gained popularity and had been used by more than 30 million people. Its features include the ability to access various services and databases to provide information such as news, weather, stock information, movie showtimes, phone book listings and sports details. It also functioned as a personal assistant, calculator and translator. SmarterChild was used as a marketing tool, and it gave rise to SmarterChild-based bots characterized as artists, musicians or movie characters. It can be said that it has paved the way for today's commercial use of chatbots. Below is an image of SmartChild (Fig. 5).

4.7 IBM Watson

This is a question-answering and decision-making support system developed by IBM. It was named after Thomas J. Watson, the de facto founder of IBM. Originally, it was

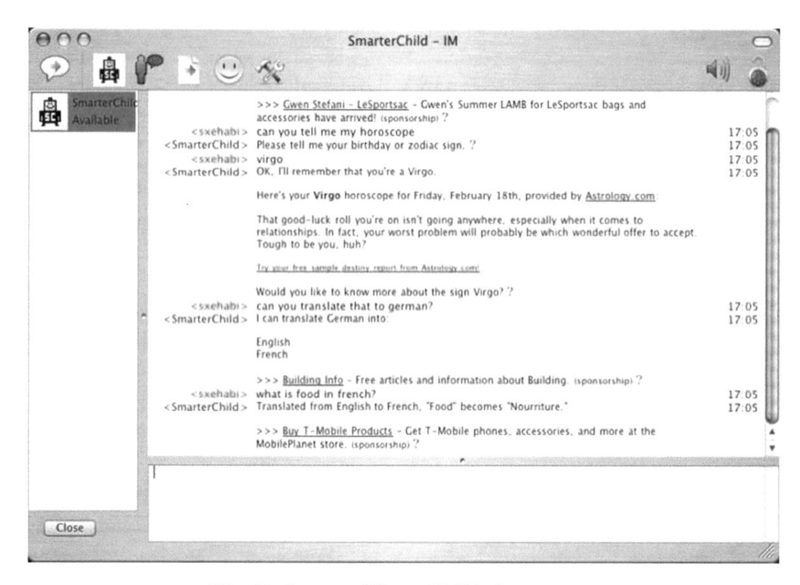

Fig. 5. Image of SmartChild chat screen.

a question-and-answer system that aimed to win on the popular American quiz show "Jeopardy!." IBM announced the completion of Watson in 2009, and it made the first public appearance in 2011. It was developed as an independent artificial intelligence and database that properly interprets the questions in natural language, and then searches for, selects, and gives appropriate answers from a vast amount of information, which is comparable to text data of about one million books. Based on these information processing technologies, Watson has been commercialized as a system which automates business processes, particularly inquiry response and knowledge search. In the process, it also provided interfaces and templates for use as a chatbot, which was introduced into the systems of various companies. Thus, it has become one of the de facto standards for business chatbots. For example, the task of instantly categorizing and organizing a vast amount of information to retrieve what is necessary is one of the areas where computers excel. Other examples include a system that presents information to call center operators based on customer's requests, or a system that selects right wines for the customers. Basically, Watson is good at presenting accumulated information according to the situation. Watson's features are also made publicly available through API. There are three major features: (1) language processing; (2) image recognition; and (3) speech recognition. It can be said that Watson is equipped with the fundamental features of today's AI.

4.8 Siri

Siri is the voice assistant system introduced on iPhone 4S (iOS 5.0) in 2011. It stands for speech interpretation and recognition interface. Siri combines speech recognition technology and natural language processing to provide various support functions such as information search, weather forecast, route search, stock information, e-mail, music, clock, calendar, alarm, timer, calculator, machine translation, reminder, reading out incoming and outgoing phone calls by connecting to the Internet and linking to various applications. For example, it responds to user's verbal requests or questions, such as "I want to know the weather forecast for the weekend" or "I want to send an e-mail to my brother." While the existing SmartChild supports text-to-speech, Siri has been implemented as a speech-recognition interface, which allows for oral commands. It was initially developed by Stanford Research Institute International for CALO, an artificial intelligence development project by U.S. Defense Advanced Research Projects Agency (DARPA), to support soldiers in the battle-field. It was a byproduct of a project that attracted huge amount of money and researchers from prestigious universities and research institutions.

4.9 Google Now/Google Assistant

Google Now was released as a virtual assistant in 2012 and was implemented as part of the Google Search application. Most applications, including existing chatbots, were passively activated by user actions and inquiries, and provided information and services in response to them. On the other hand, Google Now was able to actively provide information by recognizing repeated user actions on the device, e.g., frequently-visited pages, repeated appointments, search keywords, emails from e-commerce companies sent to Gmail accounts, in the context of location and time of day. The system uses Google's Knowledge Graph, a mechanism used to generate more detailed search results by analyzing the user's meanings and connections. Google Now was replaced by Google Assistant in 2017, and is still widely used as an AI speaker.

4.10 Cortana

Cortana is a voice assistant developed by Microsoft in 2014 which comes standard with Windows OS. Microsoft was one of the first companies to focus on office assistants, introducing the Office Assistant Clippy in Office 97 and Office 2000. Cortana is an assistant with a voice interface. By typing a question in the search box or selecting the microphone and talking to Cortana, you can use the same functions and service integration as other assistants such as Siri and Google Assistant. If you don't know what to say, you can have suggestions displayed on the lock screen or on Cortana's home screen by selecting the search box on the taskbar.

4.11 Alexa

Amazon Alexa is a virtual assistant AI technology developed by Amazon and was first used in 2014 in the Amazon Echo smart speaker developed by Amazon Lab126. It is capable of voice interaction, playing music, creating to-do lists, setting alarms, streaming podcasts, playing audiobooks, and providing real-time information on weather, traffic, sports and other news. Amazon allows device manufacturers to integrate Alexa voice capabilities into their own connected products, using the Alexa Voice Service (AVS), a cloud-based service that provides APIs to interface with Alexa. Products built using

AVS will have access to Alexa's growing list of features, including all Alexa skills. AVS provides cloud-based automatic speech recognition (ASR) and natural language understanding (NLU). Amazon Alexa's voice is generated by a Long Short-Term Memory (LSTM) artificial neural network. Users can also expand Alexa's capabilities by in-stalling skills from third-party developers. Amazon also provides a platform for virtual assistants, releasing Amazon Skill Kit (ASK) to third parties to encourage them to develop "skills" (additional functions, like apps) such as games and audio functions.

5 Algorithm

In this chapter, we summarize the algorithms for chatbots. There are four major types of chatbot algorithms.

(1) Choice type
    In this type of conversation, the user makes choices according to a predetermined scenario. Since the conversation is based on the user's choice from the answer options, the chatbot could not converse outside of the scenario. However, it can carry out a simple conversation smoothly when, for example, responding to a question.
(2) Log type
    The chatbot holds a conversation, following the context, by utilizing the accumulated conversation logs. The concept of this type of algorithm is to use a large amount of conversation data and logs to enable natural conversations. In recent years, machine learning has been applied to logs to generate more natural conversations.
(3) Hash type
    Also called dictionary type, this type of conversation proceeds based on keywords registered in a dictionary or database. It is difficult to form a conversation with words that are not in the dictionary, but if the words to be used are within the predefined range, the chatbot can respond. This would be useful if a large database or conversation data are used.
(4) Eliza type
    This type is named after Eliza, the pioneer in chatbots. It is a system where the chatbot basically focuses on listening but makes occasional responses to help keep the conversation flowing. Although the chatbot can only respond to simple questions, it can also form a conversation by asking questions like "What do you mean?" when it runs into words not found in the dictionary. The system summarizes the entered text and gives the user an impression that it has understood the content.

The algorithms of existing chatbot services and systems are basically classified into these four types.

6 Summary

In particular, we have dealt with ELIZA, the originator of chatbots, and voice recognition assistants such as Siri and Alexa, which are now widely used, and discussed various chatbots, both large and small. Algorithms for chatbots can be scenario-based with choices, hash-based with a dictionary, or log-based with existing conversation data to select responses from. Major IT companies, known as "Mega Techs," have established development platforms for chatbots and are eager to enlist users. It can be said that this is a field where expectations are rising to such an extent. Chatbots can reduce human costs and improve user convenience by being available 24 h a day. On the other hand, chatbots have some disadvantages inherent in machines, such as initial installation costs and the inability to answer questions for which responses are not pre-set. Also, since chatbot conversations tend to be monotonous and flowing, there are certain number of users who prefer to use human operators in customer consultation rooms. However, as the cost of introducing chatbots decreases and the technology enables reproduction of human touch, chatbots are expected to become more and more popular. Log-type chatbots, which are currently considered to be the most human-like chatbots, use machine learning to provide humanlike answers, but it is difficult to maintain them unless some adjustments are made. In the first place, modern technology has not yet reached the point where machines understand the meaning and context. Future research should explore the factors that make humans find humanness in the chatbot during conversation, and implement a system that helps people feel as if they were having a conversation with real human while talking to a machine. Though ELIZA was an early chatbot, it made some users at the time believe that it had a personality just by answering simple questions. Such a phenomenon called the ELIZA effect could happen in future chatbots. It is also possible to utilize the habits of users in conversation to modify the chatbot personalities, e.g., to a more reserved bot for talkative users or to a more active one for those who are quiet. In other words, we can expect a spread of chatbots that can be customized for each user. In the future, the demand for chatbots with personalities will increase.

References

1. Adamopoulou, E., Moussiades, L.: An overview of chatbot technology. In: Maglogiannis, I., Iliadis, L., Pimenidis, E. (eds.) AIAI 2020. IAICT, vol. 584, pp. 373–383. Springer, Cham (2020). https://doi.org/10.1007/978-3-030-49186-4_31
2. Cahn, J.: CHATBOT: Architecture, design, & development. University of Pennsylvania School of Engineering and Applied Science Department of Computer and Information Science (2017)
3. Shawar, B.A., Atwell, E.: Different measurement metrics to evaluate a chatbot system. In: Proceedings of the Workshop on Bridging the Gap: Academic and Industrial Research in Dialog Technologies. 2007
4. AbuShawar, B., Atwell, E.: ALICE chatbot: Trials and outputs. Computación y Sistemas **19**(4), 625–632 (2015)
5. Skjuve, M., et al.: My Chatbot companion-a study of human-Chatbot relationships. Int. J. Hum.-Comput. Stud. **149**, 102601 (2021)
6. Sharma, V., Goyal, M., Malik, D.: An intelligent behaviour shown by Chatbot system. Int. J. New Technol. Res. **3**(4), 263312 (2017)
7. Natale, S.: If software is narrative: Joseph Weizenbaum, artificial intelligence and the biographies of ELIZA. New Media Soc. **21**(3), 712–728 (2019)
8. Switzky, L.: ELIZA effects: Pygmalion and the early development of artificial intelligence. Shaw **40**(1), 50–68 (2020)

9. Adibe, F.O., Nwokorie, E.C., Odii, J.N.: Chatbot Technology and Human Deception. ISSN 1119–961 X: 286. Author, F., Author, S.: Title of a proceedings paper. In: Editor, F., Editor, S. (eds.) CONFERENCE 2016, LNCS, vol. 9999, pp. 1–13. Springer, Heidelberg (2016)
10. Park, H., Kim, H., Kim, P.-K.: Development of electronic library Chatbot system using SNS-based mobile Chatbot service. In: The 9th International Conference on Smart Media and Applications. 2020. 9th International Proceedings on Proceedings, pp. 1–2 (2010)
11. Sedoc, J., et al.: Chateval: a tool for Chatbot evaluation. In: Proceedings of the 2019 Conference of the North American Chapter of the Association for Computational Linguistics (Demonstrations) (2019)

With Post-internet Society, *The Third Agent*

Takashi Shimizu(✉)

Graduate School of Information Sciences and Arts, Toyo University, Tokyo, Japan
t_shimizu@toyo.jp

Abstract. In the increasingly ubiquitous world of IoT, what is the "medium" that brings about many workings? The theme of this paper is the search for things that will alleviate excessive interference and act as arbiters for the interactions between things and people, people and people, and things and things.

Keywords: Third agent · Medium · Object · Entzug · Smart contracts

1 Flying a Kite

In anthropologist his book Making, Tim Ingold undertakes a very interesting study of the act of "flying a kite." When flying a kite, a person pulls at the kite with a nylon string in his or her hand and runs. The kite ascends rapidly into the sky and at times pulls back at the person intensely. Ingold describes this as follows:

> Here we have a person running; and there a kite flying. They are connected by a thread. So we have an interaction between a human being and a material object. At one moment the flyer makes the running while the kite is dragged along behind, but at the next it is the kite that pulls at the hand of the flyer, as if struggling to get away...For if the kite is to act on the human flyer as well as the flyer to act on the kite – if, that is, the kite-object is to join the dance – then it must be endowed with some kind of agency. So where did this come from? So long as the kite remained indoors, it had been limp and lethargic [1].

Ingold says that the answer is that it is being blown by the air. The performance of kite flying is accomplished through the three-way synergy of the flyer, the kite, and the air. That is where the "dance of agency" is born. He borrows this expression from Andrew Pickering, who describes "a state in which a person and a non-person thing take turns dominating one another, and interacting with the material world while ebbing and flowing with one another." The kite-flyer and the kite surely have that kind of competitive relationship between each of their energies, but the air acts as a third party that causes the two-way relationship to materialize. It is not the case that the person has some kind of active intention and tries to influence the wind. The wind and the air currents are too strong, and the effects of a person clutching a nylon string are nearly non-existent. Ingold calls the air in this situation a "medium (Fig. 1)."

The essence of the operation of the third aspect is in transcending the rise and fall, the active and passive relationship of the flyer and the kite, and the changes in that

T. Matsuo et al. (Eds.): AIMD 2019, LNNS 677, pp. 71–79, 2023.
https://doi.org/10.1007/978-3-031-30769-0_7

Fig. 1. Synergy of the flyer, the kite, the air. (https://media.baamboozle.com/uploads/images/88460/1605261248_836016)

relationship, and staying put. Ingold also finds this kind of relationship in the three-way relationship between clay, hand, and potter's wheel, with the hand trying to give shape to the clay, the clay showing suitable resistance to the hand, and the wheel enduring this rivalry between the other two and providing the essential space for this rivalry to occur (Fig. 2).

Fig. 2. Potter's wheel (https://myfaithmedia.org/wp-content/uploads/2016/02/iStock_000046898796_Medium.jpg)

2 Zuhandenheit and Vorhandenheit

There is a reason for bringing up this theme here even further, dealing with the competitive relationship between the person and the non-person thing, the subject and the object, and the third agency behind the scenes—related to things called "instruments" in general. The reason is that in no other period more than today has the relationship between instruments and people been universally questioned more. We are trying to live in a world of the IoT (Internet of Things), and an environment in which all the things that are components of our everyday lives are connected to the internet. That is, all things are being made into instruments through the medium of computers.

Our bodies are no exception in their positioning within these circumstances. There is no doubt that soon enough, it will be normal for health status information to be continually sent to one's doctor using chips implanted throughout the human body. Not only physical information but information about our work and consumption behavior and activities will be easily sent and received. If this kind of society goes one step astray, it could turn out to be the ultimate surveillance society. To turn all of our things into instruments that are convenient for us is to make it possible to retrieve information sent from them using technology, no matter where they are. However, this is prone to lead into turning others into such things for us, and us into such things for others.

Whether a thing or a subject, the competitive relationship of power between the principal and auxiliary is thus extremely ubiquitous. However, can this situation of excessive mutual relationships only foretell a dystopian future that grasps the situation as solely a game of competing with one another for dominance? It is important to take a moment to consider the fundamental aspects surrounding the establishment of instruments and things as in Ingold's example above to further consider the third agency that acts in the background of this two-way relationship, while also transcending that relationship and causing their relationship to function. In the increasingly ubiquitous world of IoT, what is the "medium" that brings about these workings? A major topic for us is the search for things that will alleviate excessive interference and act as arbiters for the interactions between things and people, people and people, and things and things.

Let us now introduce more philosophical references regarding instrumentalism. These will be Martin Heidegger's famous instrument analysis and Graham Harman's discussion that now leads object-oriented ontology (OOO). In Heidegger's Being and Time, the instruments that we use as writing instruments and for work define the way we discover the world. Each of those instruments are linked to all other instruments through some kind of purpose. Roads are instruments for people to walk on, or for cars to drive on; umbrellas are instruments for weathering the rain; and umbrella stands are instruments for holding umbrellas; in this way, everything is linked.

Heidegger is a philosopher who said that (1) all things are instruments that have some purpose and (2) these things all form a network of links. He predated IoT and spoke in a much more ubiquitous form. He is also well known for establishing two classifications of the states in which these instruments appear before us. One is readiness-to-hand (Zuhandenheit), and the other is presence-at-hand (Vorhandenheit). When one masters the use of an instrument such as a hammer, the instrument is said to have a readiness-to-hand. When it is in a state of readiness-to-hand, the instrument functions harmoniously in instrument linkages, but it is not especially deliberately topicalized. However, the instrument can be removed from that functional context and seen observationally. This materialization is presence-at-hand.

When an instrument is in the state of presence-at-hand, it is not necessarily showing us its entire existence. It is only showing us a limited aspect of what it contains. What about when it is in a state of readiness-to-hand? Heidegger's work certainly placed more emphasis on instruments in a state of readiness-to-hand rather than presence-at-hand. Furthermore, pragmatic interpretations of Heidegger's philosophy, frequently seen in the English-speaking world, often conflate readiness-to-hand with instruments being freely used in a practical way.

Heidegger's idea of the network of instruments itself is perceived as like the instrumentalism of pragmatists such as Dewey and others. Things only appear as objects to a subject when they are used as instruments. Heidegger's analysis of instruments clarified that these appear before people as the world itself and as similar things. In points (1) and (2) above, Heidegger presents a pioneering point of view, even from the perspective of today's information technology society. However, in his pragmatic interpretations, subject and object do not even make a competitive relationship of energy and only see a one-sided expansion of the subject's active nature. It may be difficult to show the way to an ideal relationship between people and instruments or things based on these kinds of interpretations.

3 "Useless Uses"

Graham Harman fundamentally sympathized with Heidegger, while advancing an objection to such interpretation. According to him, such advocates placed too much emphasis on the active nature and expansion of subjects toward things, and at times concluded by saying "what humans encounter are not hammers or electrons or dolphins, but themselves." However, the important piece of Heidegger's readiness-to-hand argument was originally that instruments, and furthermore things, have an existence that is withdrawn (Entzug) from all human operations. Harman says the following, while quoting the words of Heidegger himself:

> The peculiarity of what is proximally ready-to-hand is that, in its readiness-to-hand, as it were, withdraw in order to be ready-to-hand quite authentically." In short, insofar as the tool is a tool, it is quite invisible. And what makes it invisible is the way that it disappears in favor of some that it serves [2]:

When an *instrument* is freely used for some purpose, that practice ends up obscuring the *instrument*. In free practice, the *instrument* is not intentionally topicalized, but without being inferior to presence-at-hand, it renders the instrument or thing invisible. Conversely, from here on, Harman weakens the workings of the subject to the absolute minimum and reconstructs the theory of instrument linkage.

> For Heidegger tools do not exist as isolated entities. Indeed, their very contours are designed with other entities in mind:" A covered railway platform takes account of bad weather; an installation for public lighting takes account of darkness, or specific changes in the presence or absence of daylight—the 'position of the sun'." Instead of thinking that extramental reality is founded on what appears to consciousness, we must join Heidegger in concluding opposite… [3].

An instrument in a readiness-to-hand state is not consciously topicalized when it is freely handled, but if it breaks or there is an accident, it is even more strongly noticed. Perhaps we first realize that the earth is something that supports us when we experience it shaking in an earthquake. Things or instruments withdrawing from us would be the so-called "play" portions of instances when they function effectively. In the above citation, when "a covered platform considers bad weather," that "consideration" does not refer to the fact that some designer somewhere "considered" the bad weather of some stormy night. It means that between the "covered platform" and the bad weather, there was "play" that made it all right even if the two were freely acting on one another.

In the first example of kite-flying, the person who held the nylon string and tried to raise the kite into the air did not directly try to influence the "air" or the atmosphere. When those things are not functioning, things and instruments carry out their "useless uses." This is because the linkage of instruments comes about through the "play" of such withdrawn parts. "Thirty spokes connect the hub. In this space of nothing, is the use of the car." (A car's wheel is made from a collection of 30 spokes in one hub. The car's role is played from that space where there is nothing.) These are the words of Lao Tzu, but perhaps we could also say that all things and instruments have an aspect that acts as this kind of "hub." When thinking about the linkage between instruments as this kind of structure, it seems that even if the subject is already removed from the trigger of relating to things or instruments, that network would remain. From this point of view, we need to further grasp again the question of "medium." Additionally, we should also examine whether the two-way relationship between a subject and a thing is simply useless at this (Fig. 3).

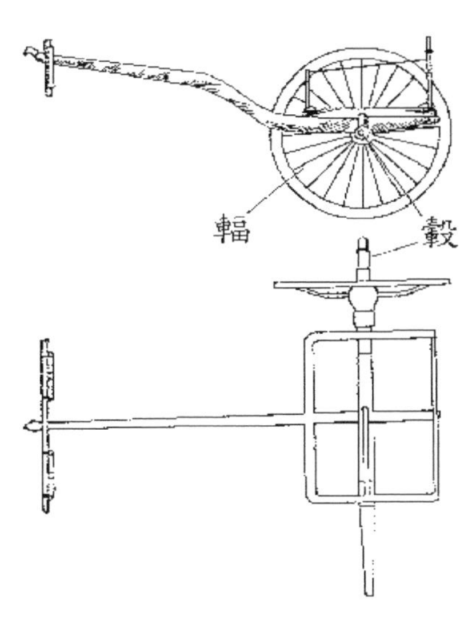

Fig. 3. "Thirty spokes connect the hub. In this space of nothing, is the use of the car." (Lao Tzu) (https://pic.pimg.tw/combohuang/1572356279-4119519387.png)

4 Where There are Relationships, There are Objects

All things and instruments are linked to form this world. However, within that linkage, there are parts that have withdrawn from the network, paradoxically bringing the network into existence. However, in reality, in two-way relationships between things and things, and between subjects and things, the two parties invade one another and interfere, and withdrawal is always threatened. The two states of a relationship, or the withdrawal of the thing, appear here as things that contradict one another. This is a choice between two things: a relationship, or a thing as withdrawal. To put it another way, we may be able to say that it is a choice between a relationship and the agents that form a relationship.

However, along with Harman, it may be possible to advance an objection to this interpretation of such two-way relationships. When a thing, or an agent, is considered to have withdrawn from a relationship, that relationship is external as seen from the point of view of the agent, and it may be supposed that the withdrawal occurs within the thing or agent. However for the thing or agent, there is also an internal relationship that comprises the thing itself. To be concise, at least internally, the thing or agent is unable to truly escape relationships.

Restoring a thing or agent to a relationship with the things of which it is comprised internally is inevitable, as with restoring the thing or agent to an external relationship containing that thing or item and another thing or item. This is because it is the restoration to a relationship, and not a departure from various things or items as they are.

What kinds of things are conceivable now for withdrawal? A two-way competitive relationship between a thing or agent A and a thing or agent B ultimately stops at being an external relationship and is something not withdrawn. However, now, what happens if we suppose a thing or agent C that further encompasses the relationship created by thing or agent A and thing or agent B?

This time, the relationship between a thing or agent A and a thing or agent B changes while they compete with one another. They do not necessarily stop at the relationship that composes thing or agent C, but in those changes, thing or agent C is considered to be something that "withstands." In the first example, this would certainly be the medium of the "air" or "atmosphere" as a third agent that appears in the competition between the two energies of the kite and the person. Since the external relationship of thing or agent A and thing or agent B is of a different class for thing or agent C, a "play" section is established in the three-way relationship. In this case, a thing or agent C is withdrawing from the perspectives of thing or agent A and thing or agent B. Then, from the perspective of thing or agent C, thing or agent A and thing or agent B are also real objects that are withdrawn.

Harman says that "where there are relationships, there are objects [things]," but the only difference between relationships and things or agents is the position from which one looks at the same thing. When standing in this point of view, the previously mentioned two choices are erased. The important thing here is not only that things or agents and relationships are regarded as the same things but also that the act of understanding one from the foundation of the other is itself repudiated. Harman thinks of objects (things) as nested and mutually encompassing to the end, and in that, all things or agents earn the same position as the third agent in the thing or agent C mentioned above.

Furthermore, because of this, even while things or agents have withdrawn from each other, they are thoroughly linked. Like Harman, we must not fail to understand that the linkage of instruments has third agents as joint sections in this sense. In order to manifest this agency of a three-way struggle, the first thing required is the competition between energies in a two-way relationship and the changes therein—the state of a kite and a person pulling at each other fiercely.

5 "Medium" for Our Society

When a bird beats its wings strongly and heads for the sky, there is competition between the energies of the bird and the environmental world. However, beyond that, there is a medium that further engulfs that two-way relationship. When the bird rides the air currents, it makes almost no idle movements. The same goes with a fish swimming in the ocean currents. Incidentally, when taking into account the state of this kind of three-agent situation in nature, how would our now-ubiquitous information environment be perceived? (Fig. 4).

Fig. 4. Gliding seagull and Medium

The issues with the discussion on IoT are that all things are grasped as information, that a society has arrived that uses this as an instrument, and that it has become increasingly possible to grasp our bodies, ourselves, as things or instruments from the perspectives of other people. That situation is, in all respects, mutually encompassing; Internet displays and the real world repeatedly encompass and invade one another. However, have we not come to understand this situation merely as an extension of a competitive relationship between the energies of a principal and an auxiliary? In that case, what computerization brought about was only an increase of the workings and surveillance of other people and things, and a society where they are mutually and eternally attempting independent and active behaviors in such a context.

It is not that this two-way energy competition is unnecessary. Rather, in addition to that relationship, what is needed here is for us and for things to be able to take on the third-agent position. What kinds of methods are possible for exploring that social system construct in the current information environment?

The first conceivable thing is for us to withdraw ourselves from the environment into the position of the third agent. By refraining from excessively exposing ourselves to the information environment, and strengthening privacy functions, it is thought that this situation will be realized. There are already many arguments being made for recovering from "connection fatigue." However, in this case, it is not believed that something will be created for this purpose that will function as the "medium" or a third agent that will further encompass two-way relationships as competitions of energy.

Therefore, the realistic assumption is to keep competitions of two-way energy in a competitive state as much as possible and create stability by substituting this third agent function, that is indispensable in the natural world, with things. This will create a so-called "medium" for our society and alleviate the excessively conflicted current situation.

On this point, some technologies have already been created that can have a large effect on society. Blockchain technology is currently primarily used to transfer cryptocurrencies. However, soon its application may bring smart contracts within reach, that is automating sales contracts and all other types of contracts with programs that do not have transparent intentions about the pursuit of certain interests. Until now, a trustworthy third party was needed to secure the execution of contracts. However, depending on their role, these third parties were influenced by their own interests, such as raising profits, and thereby participated in profit-seeking power games. By substituting them with a technology protocol, the third agent function shows a realistic and ubiquitous potential for the first time.

Because today's blockchain technology is used for tendering the easy-to-understand medium of cryptocurrencies, it is easily dragged into speculative economic activity, and as with regular currency, cryptocurrency easily contains very visible agency. However, it is doubtful that such speculative inflammatory factors would exist in smart contracts. So they would be expected to function as a "medium" for the infrastructure of social life in a dispersed society that is not under a centralized, authoritarian rule. Ethereum's 2016 presentation of a comprehensive platform for smart contracts remains fresh in mind.

The concept of contracts is one of the most important notions in politics. The social contract theories proposed by Hobbes and by Locke assumed the power games of "conflict of everyone against everyone" as the natural state of humanity. They trusted in the "greatest power" or the "greatest majority" to mediate, transferring rights of everyone to it in exchange for protection. This "conflict of everyone against everyone" is basically the expansion of the two-way conflict model to the scale of all of humanity. The efforts of modern nations continue to firmly follow the fundamentals of this line of thought, even if the interpretation of what embodies the "greatest majority" may differ. However, there is no need for a human being to always be placed in this position of the third agent, the mediator. Before long, we may gradually be able to replace many administrative jobs with technology like smart contracts (Fig. 5).

One could say that a society that has things or technology intervening in the mediation of relationships between people and things, things and things, and people and people, may be more natural in a sense than human societies. More so since the modern period,

Fig. 5. An alternative to "the greatest majority"? (https://cdn-ak.f.st-hatena.com/images/fotolife/P/PentaSecurity/20210514/20210514120548.jpg)

that only assumes that people will be in conflict with one another does not consider the perspective of things. By introducing this kind of technology into the infrastructure of economic and political activities, we should search for initiatives in a variety of forms that alleviate and disperse human-caused over-concentration and the overheating of competition.

References

1. Ingold, T.: Making—Anthropology, Archeology, Art and Architecture. Routledge, pp. 98–99 (2013)
2. Harman, G.: The Quadruple Object, Zero Books, p. 38 (2011)
3. Ibid, p. 38

Research on Digital Reading Experience with 'PAGE-Turning' Physical Feedback

Yulana Watanabe[1,2](✉) and Takayuki Fujimoto[1]

[1] Toyo University, 2100, Kujirai, Kawagoe, Japan
s4b102000049@toyo.jp

[2] Aoyama Gakuin University, Shibuya 4-4-2, Kujirai, 2100, Kawagoe, Japan

Abstract. In recent years, the Internet has become widespread, and many people have their own devices. The purpose of this research is to enhance the sense of physical reading sensation by introducing 'physical touch' to reading e-books on digital terminals, which are becoming more complex due to the various functions stored in e-books. In this paper, we focused on the 'page-turning' motion as a factor that enhances the reading experience, and conducted experiments and verifications. For the experiment, we used walnut binding, a common binding method, to verify the presence or absence of reading sensation, changes in reading sensation by the different number of pages, and convenience. The results show that the number of pages that provided the best reading experience while keeping the convenience of the digital terminal, was two sheets of paper used for 'four page-turning' operations. In addition, we conducted a verification of which position on the paper is touched and how it is turned when people turn pages for paper book reading, based on behavioral observations of 22 experiment participants. As a result, the page-turning motions were categorized into four patterns. Although there were some differences in the positions and used fingers, most people followed a pattern in which they hooked their thumbs to the bottom of the page, held the page with their other fingers, slid their thumbs under the page, and use another finger as a base point by holding the page between these fingers and turning.

Keywords: interface · Interface · Reading · page-turning · physical touch

1 Introduction

1.1 Considerations on the Complexity Caused by Digitization and Physical Touch

These days, many people possess their own devices that allow them to communicate and search for information regardless of time and place. Smartphones and iPhones are the most common examples of individual digital devices. Tablets are also gaining significant market share, and their shipments in Japan, 2020 increased by more than 30% compared to the previous year. In this way, the devices that can be operated by touching the display are rapidly becoming more popular than conventional devices such as computers, which are operated using dedicated interfaces such as keyboards and mice. The most popular

T. Matsuo et al. (Eds.): AIMD 2019, LNNS 677, pp. 80–90, 2023.
https://doi.org/10.1007/978-3-031-30769-0_8

smartphone operating system, the iPhone, had a control button called the 'home button' below the display, but this has been eliminated for the 2017 and later designs, and now it provides a front display specification. Operation by touching the display is more intuitive and it seems to eliminate complexity. However, its simplicity can also lead to confusion. The buttons prevent users from missing 'what to do next' in the operation because its presence clearly indicates the part to be touched. For example, on today's most famous device, the iPhone with full display, you can return to the Home Screen by pressing and holding the screen a little harder and swiping, after opening an application and performing any operation upon the application. On the other hand, in a conventional device with buttons, you can go back to the first screen just by pressing that part of the screen, and you don't need to do anything different from touch on the display, such as pressing and holding or swiping. Although the design with the home button, is not as simple as that of a full-screen display, the physical buttons make physical operation very simple and easy to recall. There seems to be a considerable number of people who found that this is an advantage, and the second generation iPhone SE, which has a 'home button' iPhone was the best-selling smartphone in 2021, after years of separation, surpassing all other iPhone series. Also, in recent years, iPhones have put a lot of developmental effort into their camera functions. Some iPhones are equipped with a very high-performance camera function that can capture images similar to ones by the traditional large video camera, making them more popular than those with only a regular camera function. However, it is notable that the aforementioned "iPhone SE 2nd generation" are purchased more than them.

There are several possible reasons for this, and out of them, two major factors can be considered. The presence/absence of the aforementioned 'home button' is one point. This is not because of the differences in design and function. The phenomenon seems to suggest that physical simplicity that is easy for people to deal with tangible sensation, to require "just press the button," is preferred, to 'the operation that is completed only by series of 'touch gestures', which is the result of centralizing and pursuing simplicity in appearance. The other point is that few people are able to use such a very high-performance feature with clear intentions. There are many reviews and verification videos that compare the cost effectiveness of iPhones equipped with recent advanced cameras to conventional models, and many of them conclude that they may not be necessary unless you are a video professional. In addition, the explanation of its high performance includes many technical terms, and in general, it has a strong sense of "I don't know the detail, but I hear it's amazing. Digital devices have become more complex and less easy to "master" them due to the complexity of their functions as now many people have unique devices and it resulted in the pursuing various kinds of convenience. This sense of "I don't know the detail, but I hear it's amazing" has become commonplace as various aspects of our lives have become digitalized, and it is often the case that we often use digital tools without really understanding them.

In addition, it is also the case that there will be confusion about the operability of the system, because of the tool's appearance which is nothing but simple. This is unlikely to be a major problem for younger people, who are known as digital natives. However, when they get older, the sense of 'lacking proper understanding of the tool' is more likely to become a problem. Of course, when they first-try to use a new digital tool, there would be the interests to the unknown thing and the joy of holding it in their hands. However, when they cannot understand how to use the tool regardless of effort, it leads to the feeling of rejection, and as a result, it is expected that they will not be able to enjoy the convenience of digitization. In short, we think that there is high possibility that people find inconvenience in the tools that are too highly- developed. On the other hand, certainly, there are the people who enrich their life by using new features such as the camera performance mentioned above. However, this is a reversal of reasoning. They create a purpose by relying on the performance of the device although originally the device is intended to be used for a purpose. This leads us to consider that recent digital developments have been very self-righteous, rather than evolving in line with our lives.

On the other hand, there are some things that have retained their original form and have gained a certain level of support despite the digitalization. An example is a book. In recent years, e-books have spread rapidly and there are many dedicated applications and websites, but there are still many people who prefer paper books, and the library utilization rate is increasing. This is partly due to the fact that more and more people tend to choose not to spend money for books. However, since there are many free e-books available today, it is unlikely to be the primary reason. In other words, the entire process of going all the way to a designated place such as a library, selecting a book, and flipping through the paper to read the book is tied to "reading" and has its own appeal. We believe that all of our physical interactions with the paper - actually touching it, turning the pages, putting the ones we like away on bookshelves, etc., where we can see them in our daily lives - constitute the reading experience. In this study, we focus on this kind of physical involvement and the actions to perceive the material texture.

1.2 Real World Interface and Physical Touch

With the development of computers and other electronic devices, there have been active discussions and studies on how to interact with digital; how to incorporate them into our daily life and how to utilize them. One of them is the "Real World Inter-face," which is based on our real-world representations and actions. A tutorial intro-duction to computer augmented environments (1996) states that one of the goals of Real World Interfaces is "not to create a computer that does something in place of a human, but to pseu-do-enhance human capabilities themselves." According to the study, "in traditional GUIs, the user interacts with the computer face-to-face, using a keyboard and mouse. On the other hand, the user also interacts with the real world, but the two interactions are not fused, leaving a gap, so to speak." With regard to the expressions and worlds of entering into virtual spaces such as VR, which have attracted a lot of attention these days, 'the interaction with the real world' is completely cut off. In this respect, what is pointed out by the study also applies to that there is a huge discrepancy between the way in which we interact with computers and the way in which we interact with the real world. When we use a digital device, we certainly use some kind of interface. Users have different choices such as mice and keyboards, but no matter which they use, the movement is inorganic and nothing but 'operation'. The Real World Interface combines this inorganic behavior with real-world objects to present interaction with the computer. In the similar way, "physical touch", which is the focus of this study, is quite reality-basis. We propose an interactive way to control a convenient computer by external stimuli that we usually take for granted in our daily life.

2 Purpose

In recent years, with the spread of the Internet, many devices have been stored in digital terminals. An example is a clock. Humans began using sundials as a means of telling time as far back as 4000 BC. Since then, water clocks and hourglasses have existed as a means of telling time, changing their forms while using various resources. Many of them existed only in designated locations, and people had to go there to know the time. Even when mechanical clocks were first invented in the 13th century, tower clocks and the like were common. Later, in the 18th century, watch technology developed dramatically, and watches became smaller and smaller. As a result, the pocket watch became the mainstream, and in the 19th century, the wristwatch be-came the most common as a watch. However, today, the most common action taken by many people to tell the time is not to look at the watch on their arm, but to turn on their smartphone or another device. Electronic devices always have a time display. This has led to the loss of the "clock" device, which is used only for the purpose of knowing the time. However, in recent years, there has been a trend to make these watches mobile devices. For example, an Apple watch can track your pulse and exercise to help you manage your health, or it can work as a camera button when taking photos, in addition to telling the time. Person who use a wristwatch on a daily basis, shakes the arm slightly to adjust the watch's position, then bring the wrist to eye level, even on the day when he/she forgot to wear the watch. This is a very natural and unconscious behavior that is evoked by 'checking the time'. This indicates that people who use wristwatches routinely check the time, which involves a physical action: the watch is worn on the arm, and the time can be read by bringing the arm to the eye level. We believe that human life has become flat and monotonous due to the loss of these 'physical actions based on ingrained behavioral concept'. We consider that this is also the case with different behaviors, and in particular, we examined the actions based on ingrained behavioral concept of reading. We focused on the convenience of e-books, and 'turning the page' as the biggest factor of the "physical action based on ingrained behavior concept" as to reading a paper book. Therefore, we proposed an e-book system with 'page-turning'.

3 Examination of the Page-Turning Motion

Today, there are many types of digital devices with which people can read e-books, including computers and smartphones, but in this study, we focused on tablet devices that have a large LCD screen and are operated by users' touching the screen. We con-ducted two experiments to implement a better reading experience.

3.1 Experiment 1: The Number of Pages

There are many types of books and binding techniques. In this study, we used the "case binding" method, which is the most commonly used to bind books without thread or needle, as an example. Figure 1 shows an example of a case bound book. A is the cover, and ①, ②, and ③ refer to the individual sheets of paper. As shown in the Fig. 1 (right: side view), there is a halved sheet inside of a sheet of paper folded in half in the same way. When the book shown in Fig. 1 (left: opened), pp.1,2 and pp.11,12 are printed on the paper ①. Similarly, pp. 3, 4, 9, and 10 are printed on paper ②, and pp. 5, 6, 7, and 8 are printed on paper ③. Figure 1 shows only a small number of sheets of paper as an ex-ample, but originally a book is made of many overlapping sheets of paper in the same way, as shown in Fig. 1 (left: opened). B has glue on it, which serves to connect the papers to each other, and to connect the bundle of papers to the cover. In case binding, four pages are produced on a single sheet of paper, and the cover and text are printed additionally. In order to publish a book today in Japan, the minimum number of pages is 24, or at least six sheets of paper, are required, and the book needs to be bound with a cover.

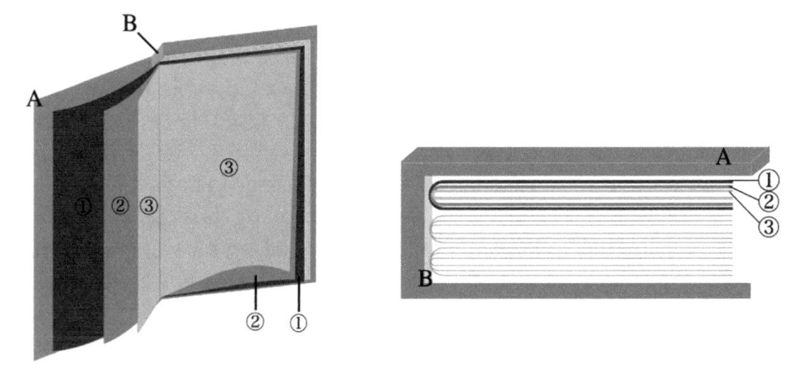

Fig. 1. A case binding book opened (left) and closed side view (right)

We considered physical sensation and ingrained behavioral concept of reading, and designed e-books to enhance a reading sensation, while maintaining the convenience of e-books. In terms of reading sensation, we focused on the importance of the physical action of turning the page. We thought it would be effective to use actual paper to enhance the perceptual reading experience. Experiment 1 was carried out as evaluation test under the condition that subjects can turn the pages of e-book by turning on actual sheets of paper. we conducted an experiment to see specifically how many sheets of paper would be effective as attachments to 'what we were aiming for in this research'. The goal of this research is to improve the reading experience by physical actions while maintaining the convenience of e-books. In the experiment, the number of sheets was reduced one by one, starting from '5 sheets (20 pages)', which is the maximum number of sheets that cannot be published as a walnut-bound book, and each was evaluated on a 5-point scale. The highest rating is 5 points and the lowest is 1 point. We examined six patterns in total, including the case where only a single sheet of paper is used as minimal attachment, in consideration of the convenience and unification of e-books, although it does not apply to the walnut-bound format.

The criteria for rating was a "reading feel" and the convenience as an e-book. In addition, a questionnaire was administered after all the number of sheets had been rated. The questions were "Do you want to use this system?", "Is it easy to use the system?", and "What do you think about the design?". We also asked the subjects which of paper or e-books they use more frequently when they usually read. In addition, to compare each case, or to summarize the whole, a free description column was provided to allow students to explain the reason for their rating and to analyze and discuss their own answer tendencies. The participants in the experiment were 27 men and women in their 20s who use computers and digital terminals on a daily basis.

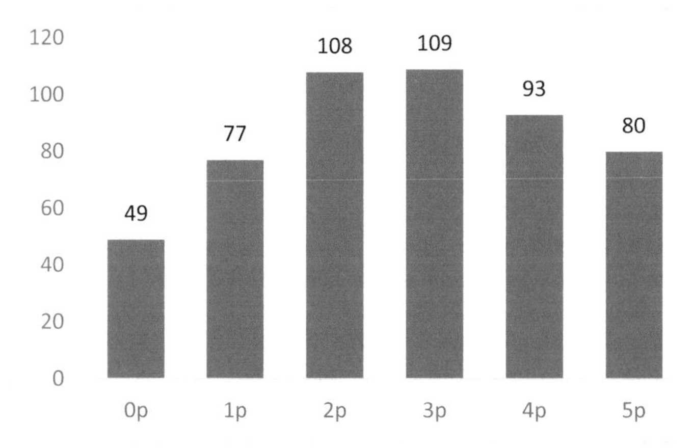

Fig. 2. Experiment.1 results

Figure 2 shows the sum of the rated points in Experiment 1. The results show that the participants were largely divided into two types: those who wanted to value the reading experience as a paper book, and those who wanted to value the convenience of a digital book. The participants who supported a large number of sheets said that the more the better because the feeling of having the next page is associated with the "reading feel." They also reported that they thought convenience was not compromised because as for '5 sheets (20 pages)', they could barely close a book-type cover. On the other hand, participants who said that they were satisfied with a small amount of sheets, rather preferred not to have sheets attached to the e-book for convenience, but they reported that in terms of the perceptual reading experience, it is certainly improved by physically touching the paper. They answered that the number of sheets was not very important, and that just turning on one page would improve the reading experience. In addition, they answered that they did not think it was a good idea to attach a large number of sheets because it may reduce visibility. Many of those who answered that 'more was better' usually read paper books, while many of those who answered that 'less was better' answered that they rather used e-books. From these results, it can be said that a large part of the preference for the number of pages de-pends on one's usual reading style.

The highest number of points was found for 3 sheets - 12 pages, but the number of points was almost the same as 2 sheets - 8 pages, therefore, an additional experiment was conducted. The method was the same as the first experiment, but the participants were narrowed down to 12 people who use both paper and e-books with almost equal im-portance in their daily reading. The results show that the optimal number of pages was 2 sheet - 4 pages.

3.2 Experiment 2: Examining the Relationship Between Finger Movements and Page Turning During Reading

In this paper, we examined how people move their fingers when reading and turning the pages of a paper book. In order to observe more unconscious behavior, we did not give the participants an outline of the experiment or any observations, but simply asked them to read for a few minutes. All the participants were filmed reading and these videos were later analyzed to see where their fingers were placed and how they "turned pages".

As a preliminary survey, we asked the participants to read e-books and analyzed their tendencies. The results showed that many of the participants were reading with their e-books tilted so that the e-book was in line with their own eyes to prevent the screen from reflecting, instead of on their desks or knees. Therefore, in an environment where there was no place to place the book, such as on a table or lap, we con-ducted a test using a common book with a hardcover cover. As a method of analysis, dots were drawn on a similarly sized piece of paper from a filmed video of the reading scene and grouped by pattern. As a result, we were able to classify them into the following four patterns. The squares are approximately the size of a finger, 1.5 cm square, and each circle is the location of a finger. The figure is an example (Figs. 3 and 4).

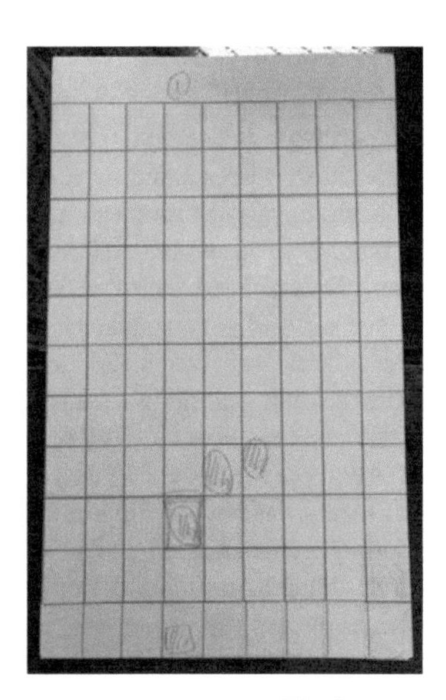
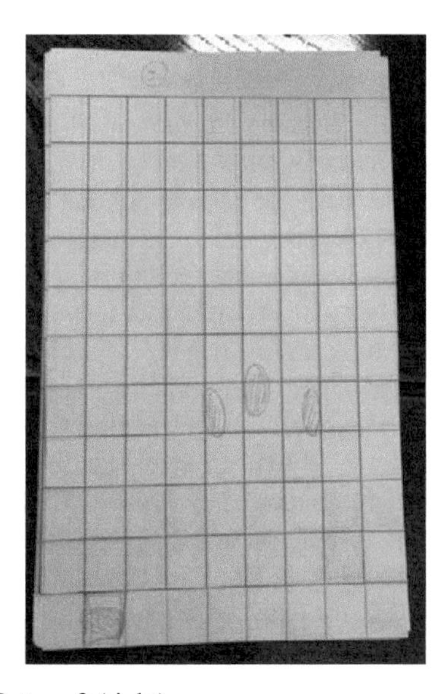

Fig. 3. Pattern 1 (left) Pattern 2 (right)

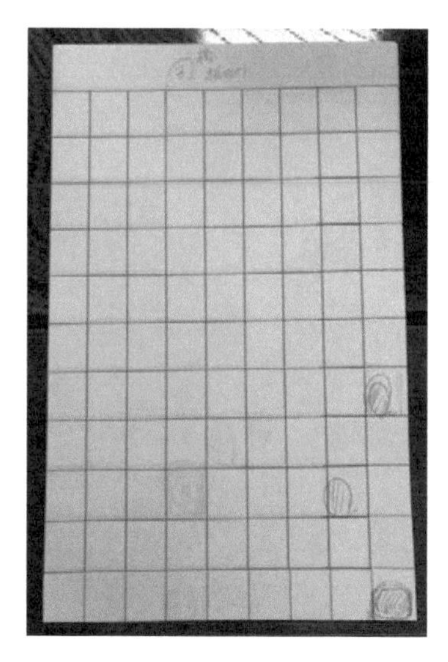
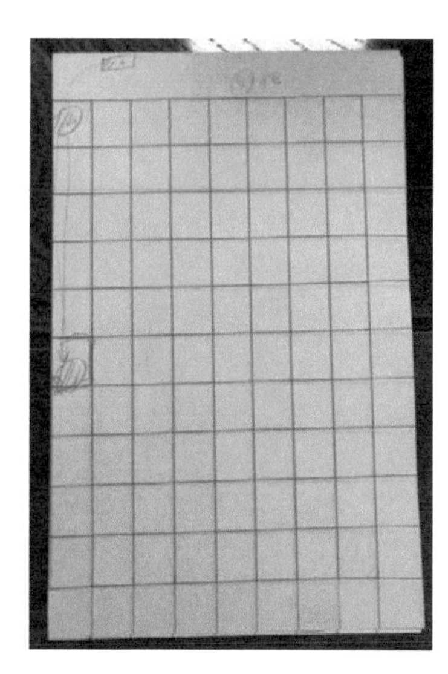

Fig. 4. Pattern 3 (left) Pattern 4 (right)

All of the experiments in this study were conducted using a type of book that can be read by turning the left page to the right to create a pattern.

Pattern 1 is placing the thumb on the bottom of the page and following the letters while holding the book with the other four fingers, mainly the index and middle fingers, on the paper. When users turn the page, they place their thumb under the paper and use their index or middle finger as a base point by holding the page between these fingers and turning it.

Pattern 2 is the same as Pattern 1, where the user places their finger on the paper to follow the text, but when turning the page, they do not "pinch" the page, but rather "slide" the page to move to the next page.

Pattern 3 uses the same finger positions and movements as 1, but instead of using fingers other than the thumb to turn the page, the thumb is hooked onto the bottom of the paper and the page is turned to the next page.

Pattern 4 is completely different from the three aforementioned patterns; a finger is placed at the top of the page and slowly moved toward the left edge as the reader progresses, and when the left edge is reached, the page is pinched between the index finger and thumb to turn the page. In this case, the pinch point was at the top left corner or in the middle of the page.

The ratios for each pattern are shown in Fig. 5.

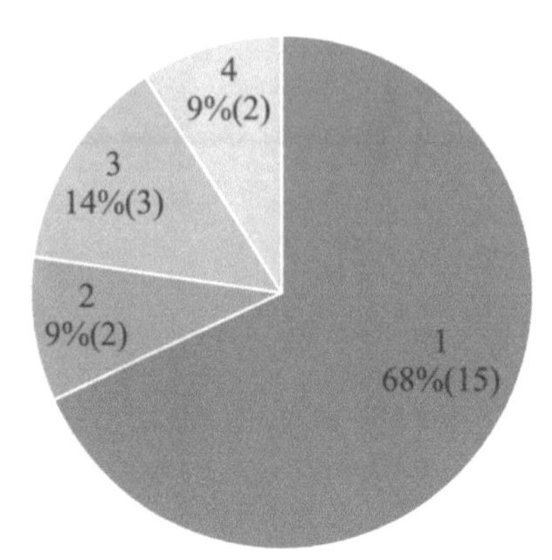

Fig. 5. Example of Pattern 1

Pattern 1 was the most common, accounting for about 70% of the responses. This means that even though there are multiple patterns of page-turning behavior, most people turn pages with their thumbs at the bottom of the page and their other fingers at the base of the paper.

Figure 6 shows an example of pattern 1. When reading, the participant placed their middle and ring fingers on the fourth square from the bottom, slightly to the right of the page in the middle of the book, as shown on Fig. 6. The index finger was placed one square below the ring finger, and the thumb was placed approximately one square below the ring finger. When turning the page, the participant lowered the position of their index finger by about one square and pinched the bottom of the page as they turned it. In this

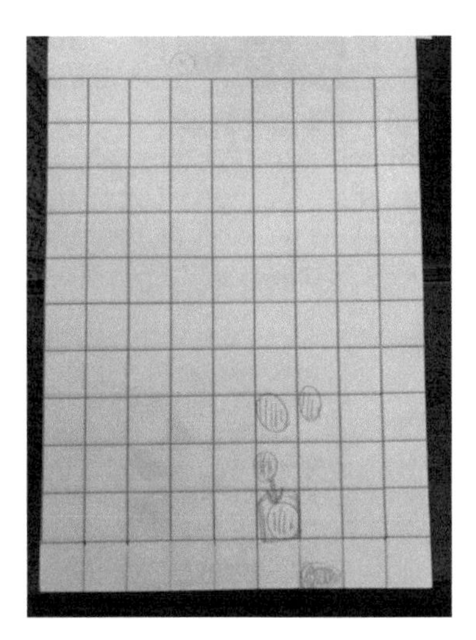

Fig. 6. Example of Pattern 1

case, we determined that the base point was the second square from the bottom and the fourth square from the right.

Similarly, further analysis of all Pattern 1 cases showed that seven people placed their fingers on the right side of the page near the center of the open book and turned the page, seven people pinched the middle of the page, and 1 person pinched the left side, the edge of the page. In all the patterns, nine of the respondents, or more than half, placed their fingers in the fourth square from the bottom and used the second square from the bottom as the base point.

4 Conclusion

In recent years, digital terminals, smart home appliances, and other highly convenient devices that have accompanied the spread of the internet have become common-place in our daily lives. Smartphones and other devices that that are owned by everyone have become so pervasive that many people say they cannot live without them. The development of these products has been remarkable, and new products with higher functions and performance are being created and released every day.

However, we believe that what humans want is something much simpler. It takes time just to master a high-performance product, and there are probably many behaviors that result from the presence of these features. We considered that there was a reversal in use and function, in which devices are now used for the sake of using them, rather than for their original purpose of performing a function. At the same time, some devices have retained their traditional forms even in today's digitalized world. For example, watches, CDs, and books. Digital devices that we always carry with us, such as smartphones, always contain the ability to see the time. This is why younger generations, who have used digital devices since childhood, are less likely to use watches. Digital watches such as smartwatches are also on the market. On the other hand, there are many people who wear analog wrist watches. Thus, there is a certain popularity of conventional products.

In this study, we focused on e-books, which have been spreading dramatically in recent years, and traditional paper books. Paper books are sometimes unsuitable for carrying around due to their weight and size. In addition, each book needs to be purchased or borrowed, and preparation is required before reading can begin. On the other hand, the device you usually use can work as an e-book, and can be used any-where and at any time although there are also dedicated devices to read e-books in the market. In spite of this, we thought that one of the reasons why paper books continue to be used is because they provide a 'sense of reading' through physical feed-back. We believe that reading books not only with the tips of the fingers, but also by moving the body and using physical movements leads to an improved sense of reading and concentration. In this paper, we focused on "page-turning" from the perspective of 'physical touch' in order to improve the reading experience of e-books. We conducted an experiment assuming that there is a system that allows users to read by turning an actual piece of paper when reading with a digital device. This experiment revealed that a four-page page-turning system can im-prove the reading experience without compromising the convenience of e-books. In addition, we clarified the position and method of touching when turning pages in paper books by observing the behavior of 22 participants in the experiment. As a result, most of the participants touched the bottom of the page and turned the page, especially more than 70% of the participants hooked their thumb to the bottom of the page and placed their other fingers on the paper as a base point.

In the future, we will consider the system design more clearly based on the results of these experiments conducted in this paper.

References

1. Okamoto, K.: Restoration of old western printed books -demonstration: repairing walnut bindings-[Japanese]. J. Tokai Area Univ. Libr. Assoc. (2), 9–15 (2004)
2. Murata, F.: A Study for the Instruction of Reading, Memoirs of Saitama Junshin Junior College, no. 3, pp. 67–73 (2010)
3. Rikihisa, Y., Moroi, K.: Cognitive mechanisim underlying reading behavior: the roles of attributional complexity [in Japanese]. DWCLA Hum. Life Sci. (45), 37–43 (2012)
4. Rekimoto, J.: A tutorial introduction to computer augmented environments. Jpn. Soc. Softw. Sci. Technol. Comput. Softw. **13**(3), 196–210 (1996)
5. Rekimoto, J.: Fusion of virtual and real: real world user interfaces: extending direct manipulation environment into physical space. Inf. Process. Soc. Jpn. IPSJ Mag. **43**(3), 217–221 (2002)
6. Kimura, A.: Tool as real world oriented interface, human interface society. J. Hum. Interface Soc. Hum. Interface **12**(2), 105–110 (2010)
7. Ikeda, Y., Kimura, A., Sato, K.: ToolDevice: real world interface using tool metaphors [Japanese]. In: Information Processing Society of Japan, IPSJ Transactions on Symposium, No.7, 2003, pp. 207–208 (2003)
8. Fujimoto, T.: Ideology of AoD: analog on digital-operating digitized objects and experiences with analog-like approach. In: 1st International Conference on Interaction Design and Digital Creation/Computing, pp. 901–906 (2018). https://doi.org/10.1109/IIAI-AAI.2018.00182
9. Fan, Z., Fujimoto, T.: Proposal of a scheduling app utilizing time-perception-reality in analog clocks. In: 1st International Conference on Interaction Design and Digital Creation/Computing (IDDC 2018) (2018)
10. Yamamoto, K., Choi, W.J., Miura, K.: The effect of tactile information on tactile impression induced by visual textures. Jpn. Psychon. Sci. Jpn. J. Psychon. Sci. **33**(1), 9–18 (2014)

Method of Extracting Community Colors with Local Characteristics in Landscape Planning: Planning the Color Scheme for the Taisetsu-Furano Route of the Scenic Byway Hokkaido

Kasai Daisuke[1](✉) and Miyauchi Hiromi[2]

[1] Tokyo Metropolitan Public University Corporation Advanced Institute of Industrial Technology, 1-10-40, Higashiooi, Shinagawa-ku, Tokyo 140-0011, Japan
kasai-daisuke@aiit.ac.jp

[2] Design Integrate Inc., Gakuen-Higashi-machi, Kodaira-shi, Tokyo 187-0043, Japan

Abstract. Sixteen years have passed since the enactment of the Landscape Law in 2005. The number of landscape administration organizations in Japan has grown to 759. Of them, 604 (as of March 31, 2020) are organizations that have formulated landscape plans, and many regions are working hard to create cities that take advantage of the characteristics of each region. Among these, color is one of the elements that seem to be the most effective for landscapes. However, few color standards provide a sense of regional characteristics, and a method for formulating color standards based on regional characteristics is yet to be established. This study is a concrete summary of developing color standards with local characteristics for the Taisetsu-Furano route of the Scenic Byway Hokkaido, which was undertaken in 2007.

Keywords: Landscape · Color Planning · Image

1 Introduction

Since the full enforcement of the Landscape Law in June 2005, residents' awareness of landscapes has been increasing. Many regions are working on urban development, taking advantage of the characteristics of each region. Among them, color is one of the elements that seem to be the most effective for landscapes.

Since 2005, Hokkaido has been working on the "Scenic Byway Hokkaido" project. People living in the region take the initiative, and businesses and governments cooperate to create unique and vibrant regions, landscapes, and attractive tourist spaces. As of the end of December 2019, there were 13 designated routes and three candidate routes for Scenic Byway Hokkaido, and approximately 440 organizations were active.

This study summarizes the process of developing color standards for the Taisetsu-Furano route, which was undertaken following the development of a color plan for the East Okhotsk Scenic Byway reported in the past [1], which takes advantage of the characteristics of the region.

T. Matsuo et al. (Eds.): AIMD 2019, LNNS 677, pp. 91–100, 2023.
https://doi.org/10.1007/978-3-031-30769-0_9

2 Research Methods

We conducted a natural environment survey and a survey on the image of residents had of their surroundings, to extract colors that make the most of the local characteristics.

In the natural environment survey, we conducted two types of surveys, fixed-point observation and moving observation, to understand the changes in color throughout the 12 months, and the colors that could be seen only in the area.

The purpose of the community image survey was to understand the kind of image residents had of the community where they lived by conducting an image questionnaire utilizing original adjectives.

It is essential to unify the region's natural environment and the image of the people who live there to extract regional colors. The colors obtained from each of these were selected as the regional colors (Fig. 1).

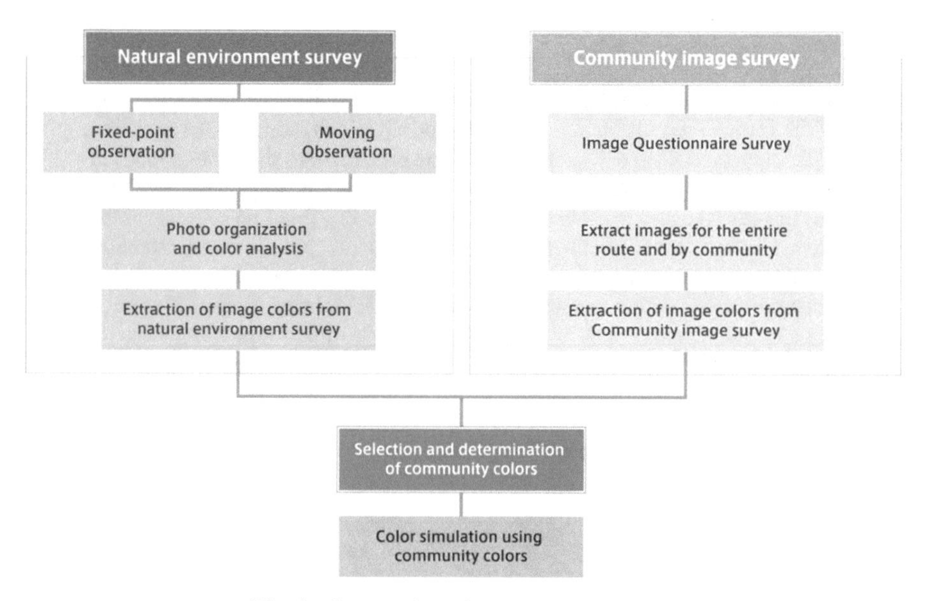

Fig. 1. Community color extraction method.

2.1 Natural Environment Survey

Fixed-point Observation. (1) Purpose of fixed-point observations. The fixed-point observation survey was conducted by taking photographs from the exact location in the same direction for 12 months, which enabled us to understand the changes in the seasons. This survey aimed to reveal the subtle changes in the landscape from season to season and to understand which places are the most beautiful in which season.

(2) Fixed-point observation method. The Taisetsu-Furano route traverses the region on a single road. Twenty-three observation points were set up at intervals of about 5 km

along the route, and observations were conducted in four areas: rural areas, hill areas, mountainous areas, and mountainous areas.

Photographs were taken every month from 10 a.m. to 2 p.m. on a sunny day from four directions (left, right, front, and back) of the observation point.

Color analysis of the photographs was conducted using Image2Assort 6.0 [2] for 10 locations in each stage that met the following six conditions:

① Many colors can be seen simultaneously.
② The image must have elements that provide a sense of perspective and three-dimensionality.
③ The composition of the entire picture should have a sense of perspective, with as many perspectives as possible from the far, middle, and near viewpoints.
④ Changes can be imagined according to time and season.
⑤ Anyone can be impressed by the beauty of the image (it is crucial to make it easy to understand as a general view).
⑥ It matches the characteristics of the area, route, and stage (Fig. 2).

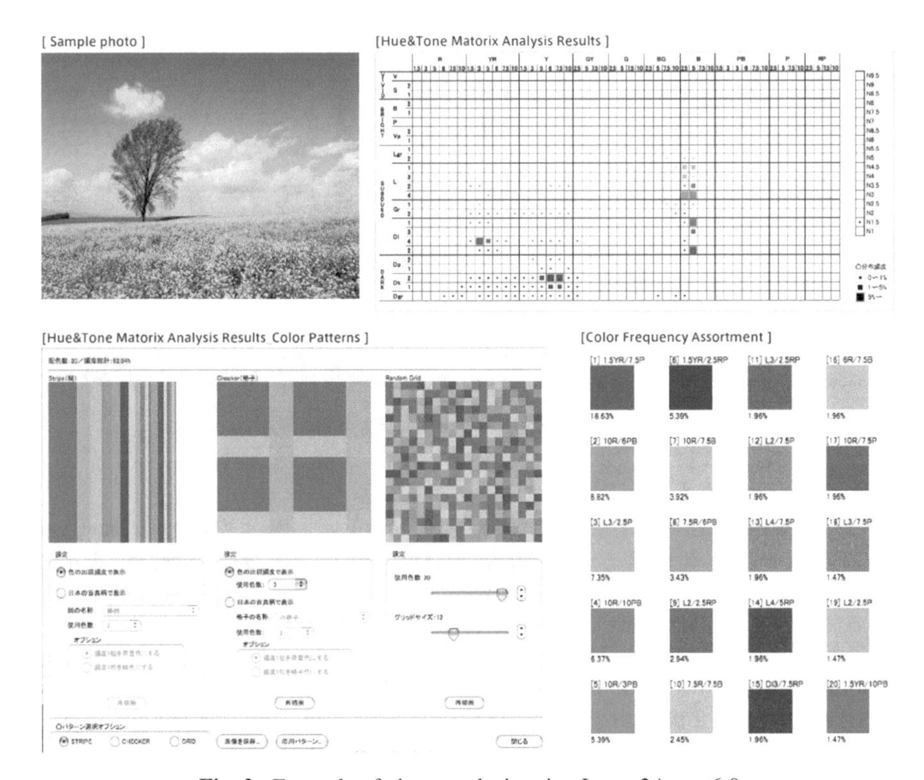

Fig. 2. Example of photo analysis using Image2Assort6.0

(3) Fixed-point observations. Photographs were taken at 10 locations over 12 months, and color analysis was performed using Image2Assort 6.0. The color area of each photo obtained from the color analysis is represented by a color bar (Fig. 3).

Fig. 3. Example of fixed-point observation results.

Moving Observation. (1) The purpose of moving observations. The purpose of the mobile observation was to identify the elements that make people want to take a "detour." The objective was to capture the entire landscape as a color image through colors that can be seen only in each location.

(2) Method for moving observations. The photographs were classified into three categories: close-up view, middle view, and far-away view. For color analysis of the photographs taken, we used Image2Assort 6.0, the same software used for the fixed-point observation survey.

(3) Moving observation results. Each photo was analyzed using Image2Assort 6.0, and nine color assortments were extracted according to the color area ratio of the photo and the image (Fig. 4).

2.2 Community Image Survey

Purpose of the Community Image Survey. The purpose of this project was to clarify what kind of image residents have of the Taisetsu-Furano route and the region in which they live.

2.3 Analysis Method

Image questionnaire survey using WAT9Analysis2.5 [3].

2.4 Target Area

A questionnaire survey was conducted in the area facing the Taisetsu-Furano route.

2.5 Questionnaire Subjects

The subjects were residents who had lived in the region for more than 10 years: 10 people per region, with no restrictions on age and gender, as they did not significantly impact the analysis of WAT9Analysis2.5.

2.6 Results and Analysis of the Community Image Survey

The image of the Taisetsu-Furano route is based on a natural image from six linguistic images: "natural," "simple," "rustic," "harmonious," "tasteful," and "simple." This can be thought of as the way people live in vast natural, open spaces. Further, with the natural image as the axis, the Taisetsu-Furano route is thought to be the images of lavender fields and sunflower fields in Furano and Biei, described as "pure" and "fairy tale," and the cultural attractions of Asahikawa, which are described as "elegant" and "enchanting" (Fig. 4).

2.7 Extraction of Regional Colors

Based on a natural landscape survey and community image survey results, basic and emphasized colors were extracted.

Base Color. The base color of the area was chosen as a subdued color as it was to be used for a large area. This led to the implementation of the extraction from the fixed-point observation of the natural environment survey, which is considered the area's foundation. The community image survey was conducted on the residents.

(1) *Extraction method of base colors.* Using Image2Assort 6.0, we overlaid 219 photos collected during a fixed-point survey of the natural environment with the colors of adjectives extracted from the results of the community image survey of residents [3]. The extracted colors were balanced on the basis of the data from all the stages to derive the base color (Fig. 5).

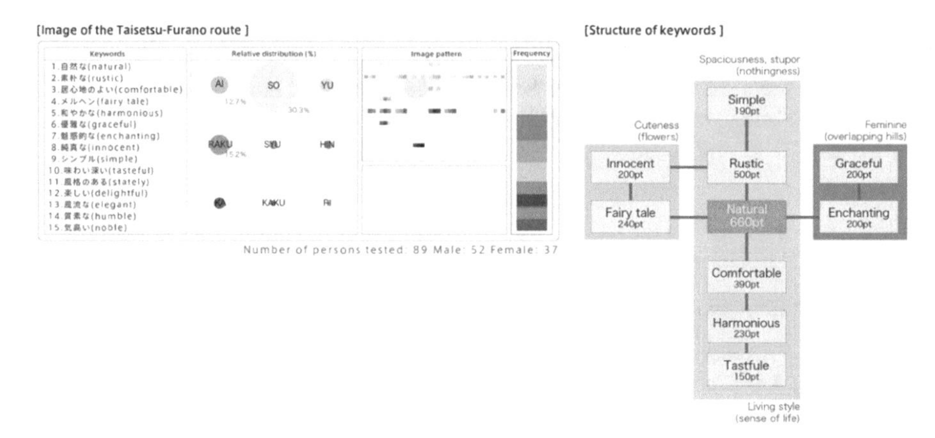

Fig. 4. Image of the Taisetsu-Furano route.

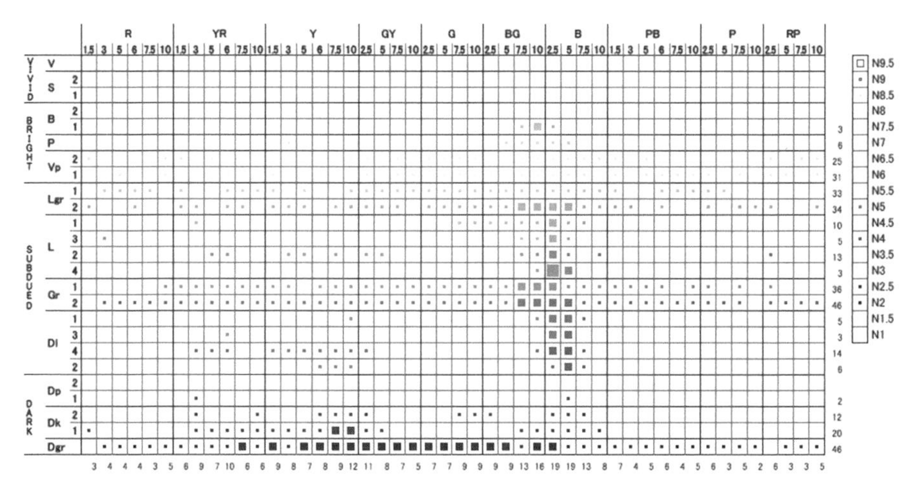

Fig. 5. Fixed-point photo analysis results (all 219 photos).

(2) *Selection of base color*. The most common colors on this route are blue-green to blue low-saturation and low-light, yellow-red to blue-green dark grayish tones, and achromatic tone. In this range, the basic colors were set to correspond to colors that would blend in with the natural environment. Eighteen colors were selected to not overlap in hue, from the concentrated irregular red to the dark grayish of blue-green and grayish of blue (Fig. 6).

Accent Color. The accent color was chosen to be slightly lighter in tone than the basic color because it is used for small areas. Accent colors were extracted on the basis of the relationship between adjectives and colors from the moving observation survey of the natural environment survey and community image survey of residents.

Fig. 6. Base colors (all 18 colors).

(1) *Extraction method for accent colors.* Using Image2Assort 6.0, we superimposed the colors of the adjectives extracted from the results of the community image survey onto 60 photographs collected during a mobile observation survey of the natural environment and balanced and extracted the emphasized colors based on data from all stages (Fig. 7).

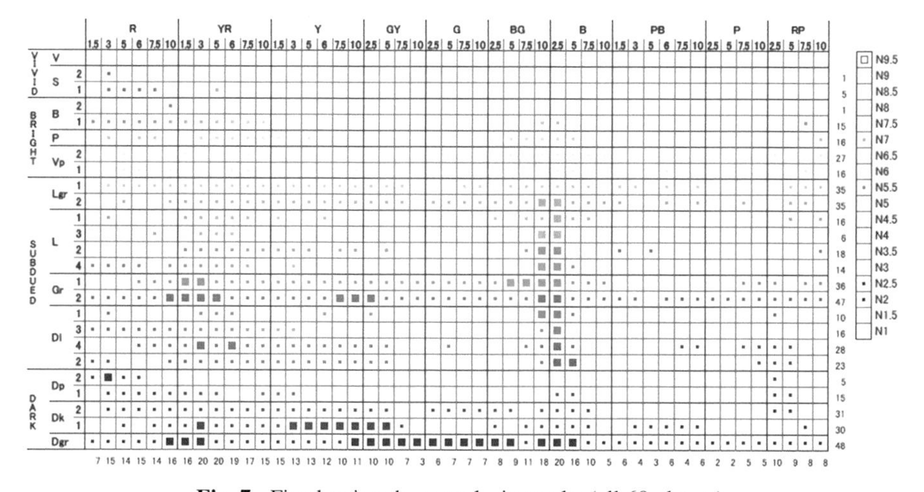

Fig. 7. Fixed-point photo analysis results (all 60 photos).

(2) *Selection of base color.* Accent colors should not only stand out but should stand out in balance with basic colors. For this reason, a total of 32 accent colors were chosen, focusing on natural colors that exist on the route, such as young leaves, flowers, sunsets, and snow, which were observed during the moving observations (Fig. 8).

Fig. 8. Accent colors (all 32 colors).

2.8 Extraction of Regional Colors

General and regional color names were used to extract basic and accent colors. By setting color names associated with the region, we aimed to make the colors more attractive to residents. The Munsell values were also set to make it easier to use the colors when installing or repainting buildings, civil engineering structures, and advertising signs.

The combination of base and accent colors is called "Natural harmonic colors in TAISETSU-FURANO." This is a combination of adjectives and nouns that evoke the natural environment of Taisetsu and Furano to make the colors themselves easier to associate with, and more familiar (Figs. 9 and 10).

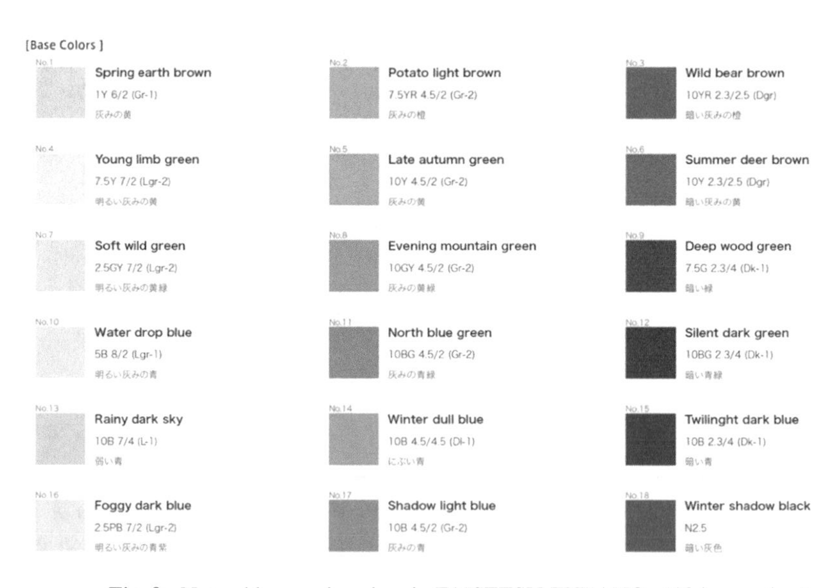

Fig. 9. Natural harmonic colors in TAISETSU-FURANO_ (18 base colors).

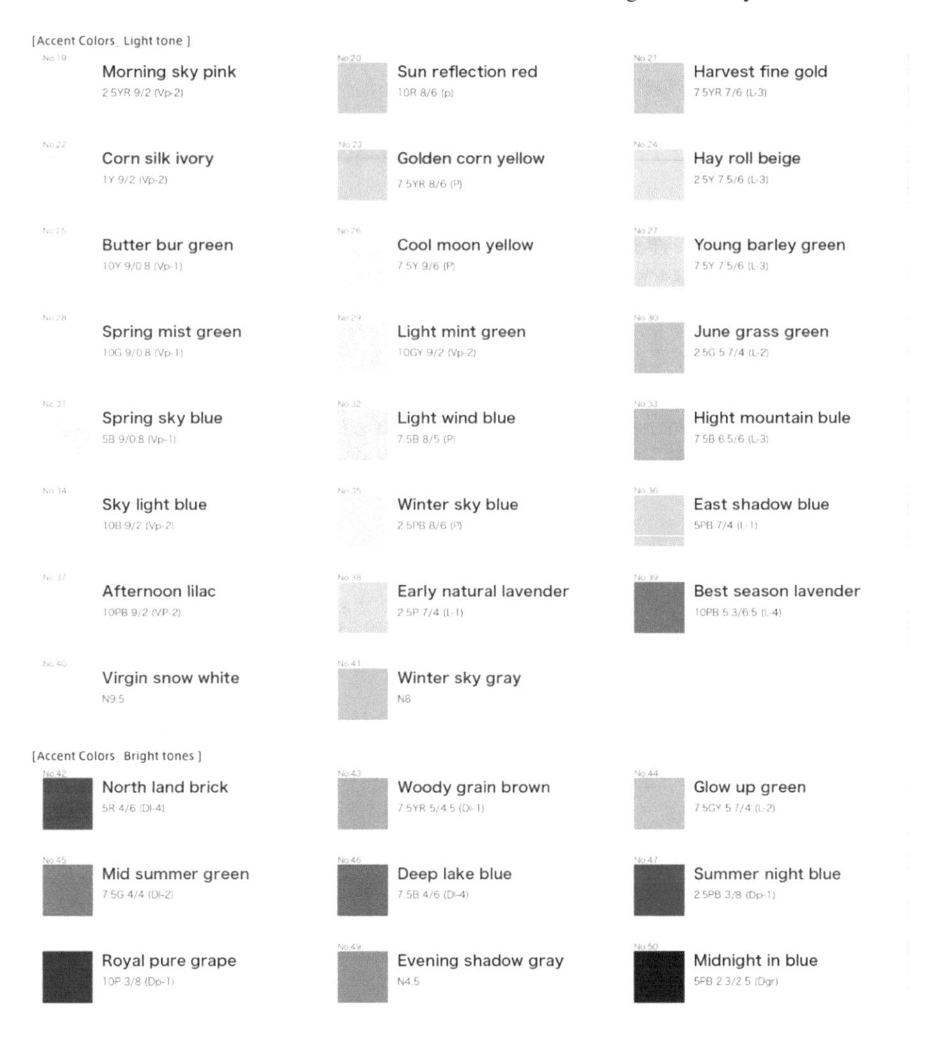

Fig. 10. Natural harmonic colors in TAISETSU-FURANO_ (32 accent colors).

3 Results

From the natural environment survey results and the community image survey, we selected 50 natural harmonic colors in TAISETSU-FURANO. In addition, the color names were chosen based on the residents' image of the Taisetsu-Furano route and the natural environment that gives a sense of the region.

4 Discussion

We compared the natural harmonic colors in TAISETSU-FURANO and the natural harmonic colors in HIGASHI-OKHOTSK selected in this study. It can be seen that the natural harmonic colors in TAISETSU-FURANO have more warm, bright, and saturated colors, overall. This is thought to be because the Taisetsu-Furano route does not face the ocean and has many seasonal flowers. By looking objectively at the environs of Hokkaido, which is richly endowed with natural beauty, we see in it various aspects and the colors woven by nature.

The 50 colors selected are guidelines to harmonize the entire landscape, not to suppress the sensitivity and sense of each local resident, but to be understood by more people and used for a few years so that it can be gradually adjusted as an image.

5 Future Work

Future work will demonstrate whether the process used in this study to select regional colors is also effective in areas other than those with abundant nature.

References

1. Daisuke, K., Hiromi, M.: Choose the right color on the east Okhotsk scenic byway. Japan Society of Kansei Engineering 10th, ROMBUNNO.12E-02 (2008)
2. Kazuhiro, O., Hiromi, M.: Development and application of color analysis software. Japan Society of Kansei Engineering 6th, p. 147 (2004)
3. Kazuhiro, O., Hiromi, M.: Word association test by 117 adjectives. Japan Society of Kansei Engineering 7th (2005)

Non-pharmacological Treatment of Dementia from the Perspective of Applied Informatics

Ken-ichi Tabei(✉)

Advanced Institute of Industrial Technology, Tokyo Metropolitan Public University Corporation, Tokyo 1400011, Japan
tabei-kenichi@aiit.ac.jp

Abstract. It has been estimated that 35.6 million people were living with dementia worldwide in 2010, with numbers expected to almost double every 20 years, to 65.7 million and 115.4 million in 2030 and 2050, respectively. Given the worldwide increase in dementia rates and the lack of a cure, non-pharmacological therapies are very important. In the past few years, many non-pharmacological therapies, including exercise, have been shown to be effective. Just as medication should be administered based on patient information, non-pharmacological therapies should not only be administered in the dark, but should also be administered with patient information in mind based on applied informatics. Thus, non-pharmacological treatment for dementia may become more effective.

Keywords: Dementia · Non-pharmacological treatment · Music therapy · Physical exercise with music

1 Symptoms of Dementia

It has been estimated that 35.6 million people lived with dementia worldwide in 2010, with numbers expected to almost double every 20 years, to 65.7 million and 115.4 million in 2030 and 2050, respectively. In 2010, 58% of all people with dementia lived in low or middle income countries, with this proportion anticipated to rise to 63% in 2030 and 71% in 2050 [1]. In Japan, dementia affects one in ten people over 65 years of age and one in three to four people over 85 years of age. The prevalence of mild cognitive impairment, a condition in which cognitive functions, such as memory, decline to an extent that is more than age-appropriate, but not to the point of dementia, is estimated to be 15%–25% among the elderly aged 65 years and above. Dementia is one of the most serious problems in society today due to accidents caused by driving, wandering, and the burden of care.

Dementia is a syndrome that is usually caused by chronic or progressive brain diseases and consists of a number of higher brain dysfunctions such as memory, thinking, disorientation, comprehension, calculation, learning, language, and judgment. Just as abdominal pain can be caused by stomach cancer or appendectomy, or headache can be caused by brain tumor or subarachnoid hemorrhage, similarly, dementia is a syndrome

T. Matsuo et al. (Eds.): AIMD 2019, LNNS 677, pp. 101–110, 2023.
https://doi.org/10.1007/978-3-031-30769-0_10

that can be caused by a variety of diseases. Dementia differs from developmental disorders in that cognitive functions that once were normal are persistently impaired due to acquired brain damage [2]. Although normal aging also causes higher-order dysfunction, dementia differs in that the patient is not aware that he or she is forgetting things, which thereby interferes with daily life.

The symptoms of dementia can be divided into core symptoms and behavioral and psychological symptoms of dementia. Core symptoms result from brain damage and include memory impairment, in which the person is unable to remember new things; disorientation, in which the person is unable to recognize the time, place, and people around him/her correctly; and executive dysfunction, in which the person is unable to carry out plans. On the other hand, behavioral and psychological symptoms are secondary to core symptoms, and include insomnia, wandering, hallucinations, and delusions. Behavioral and psychological symptoms often incur a greater burden to caregivers than do core symptoms.

Dementia can be divided into two types: degenerative dementia, caused by degeneration of the brain parenchyma, and vascular dementia, caused by damage to the cerebral blood vessels. Causes of dementia include progressive supranuclear palsy, cortical basal ganglia degeneration, lethargic granular dementia, neurofibrillary dementia, Huntington's disease, vascular dementia, prion diseases, and medical diseases. In this article, I will briefly describe four representative dementias: Alzheimer's disease, Lewy body dementia, frontotemporal dementia, and vascular dementia. Alzheimer's disease is the most common cause of dementia, followed by vascular dementia and Lewy body dementia (Fig. 1; [3]). Alzheimer's disease, vascular dementia, and Lewy body dementia are sometimes referred to as the three major dementias.

Alzheimer's dementia often begins with near-term memory impairment, followed by disorientation, executive dysfunction, and visuospatial impairment [2]. It is characterized by impaired near-term memory based on retention time and impaired event memory based on content. In contrast to proximal memory, remote memory is relatively well preserved. Disorientation often first affects one's sense of time, followed by place and person, while executive function impairment interferes with daily tasks such as work and housework. Visuospatial impairments make it difficult to copy shapes and can result in patients getting lost in their neighborhood. As the disease progresses, general intellectual functioning is impaired, and the patient gradually loses cognition of his or her surroundings, becomes unable to communicate, and eventually becomes mute.

Lewy body dementia is associated with impairments in attention, executive functioning, and visuospatial cognition, with memory impairment that may remain unremarkable in the early stages of the disease [2]. Parkinsonism, gait disturbances, autonomic symptoms, olfactory disturbances, visual hallucinations, delirium, sleep disturbances, and psychiatric symptoms are more common in early Lewy body dementia than in early Alzheimer's disease.

Frontotemporal dementia represents a progressive form of dementia characterized by behavioral abnormalities, psychiatric symptoms, and language disorders [2]. It develops latently, characterized by decreased spontaneity, disinhibition, personality change, and abnormal behavior; after the patient exhibits homophobic behavior, mental functions become severely impaired, and the patient becomes immobile, mute, and bedridden.

Vascular dementia is a dementia caused by cerebrovascular disorders such as cerebral infarction, cerebral hemorrhage, and subarachnoid hemorrhage, accompanied by depression, decreased spontaneity, and executive dysfunction [2]. Forgetfulness is often mild and subjective. It is also associated with aphasia, visuospatial impairment, and motor impairment, depending on the localization of the affected cerebrum. Symptoms tend to fluctuate from day to day owing to the inadequate circulation of cerebral blood flow.

2 Dementia and Non-pharmacological Treatment

There is no core cure for dementia [4]. There are several reasons for this, including the delay in elucidating the pathogenesis of the disease and the lack of a precise test index. In addition to pharmacological treatment with anti-dementia drugs that slow down the progression of symptoms, non-pharmacological therapies such as cognitive function training, exercise therapy, reminiscence, and music therapy are more widely used than in other diseases (Table 1).

Exercise has been reported by many observational studies to be associated with a lower incidence of dementia. Intervention trials of physical activity in older adults without dementia or with mild cognitive impairment have reported that it reduces cognitive decline, and active inclusion of exercise is recommended.

Music therapy has the potential to improve the behavioral and psychological symptoms of dementia. In addition to music therapy, cognitive behavioral therapy may be effective as a non-pharmacological treatment for anxiety [5]. Furthermore, group activities, tactile care, and massage have been shown to be effective as non-pharmacological treatments for agitation. Moreover, the use of social support and reminiscence techniques are effective non-pharmacological treatments for depressive symptoms.

3 Dementia and Musical Performance

Few studies have examined whether the ability to perform is preserved after the onset of dementia, or whether musical performance is effective in preventing or inhibiting the progression of dementia.

Cowles et al. [6] reported on a patient with moderate Alzheimer-type dementia (SL) without aphasia. Previous studies reported that Alzheimer's patients continued to perform familiar songs after the onset of the disease, but there were no reports of dementia patients being able to perform novel songs. In contrast to deficits in replay and recollection of memory tests (words, stories, environmental sounds, instrumental sounds) and deficits in remote memory (famous faces, autobiographical memories), SL showed retention of songs at 0 and 10 min. SL had better memory for the melody of songs than for lyrics or poems.

Cho et al. [7] documented a patient (JK) who first learned to play the saxophone after being diagnosed with frontotemporal dementia. Some patients with frontotemporal dementia show artistic and musical abilities. However, how patients with frontotemporal dementia can learn to play an instrument after the onset of the disease has not been reported. JK learned a repertoire consisting of 10 folk songs over a period of 3 years.

Furthermore, his saxophone skills were superior to those of the other students in his class.

In contrast to the proximal memory of Alzheimer's disease, remote memory is relatively preserved, a characteristic also reflected in performance. This is one reason for which music therapy often involves singing music that was popular when the participants were young as a form of reminiscence. However, it is generally accepted that it is difficult for Alzheimer's patients to memorize and perform new music, as in the above study, and it is questionable whether patients with Alzheimer's disease, which is a degenerative disease that progresses slowly, can maintain their ability until the end, as in the case of the patient in Cowles et al. [6]. Therefore, it is necessary to verify whether the degenerated brain regions of the patients reported by Cowles et al. differ from those of the general Alzheimer's disease population. In contrast, frontotemporal dementia does not cause memory impairment in many cases, so it is considered possible for patients to learn and perform new music. However, as the disease progresses, the ability to memorize and perform new music may be limited to a certain period of time due to the decrease in spontaneity, depression, personality change, and abnormal behavior.

Verghese et al. [8] assessed whether increased participation in leisure activities could reduce the risk of dementia. Among leisure activities, reading, playing board games, playing musical instruments, and dancing were associated with a lower risk of dementia.

Balag et al. [9] examined the association between playing a musical instrument and whether or not a person developed dementia in twins. Although there is growing evidence that playing music is beneficial for cognitive development and health in young adulthood, it is not known whether playing a musical instrument inhibits the development of dementia. The results showed that playing a musical instrument was significantly associated with a lower incidence of dementia. This result supports the effect of music as a protective factor against dementia.

Doi et al. [10] tested the hypothesis that a long-term leisure activity program would be more effective than a health education program in reducing the risk of cognitive decline in older adults with mild cognitive impairment who are at high risk for dementia. A total of 201 Japanese adults with mild cognitive impairment (mean age 76.0 years, 52% female) participated in the study. Participants were randomized to one of two cognitive leisure activity programs: dancing ($n = 67$), playing a musical instrument ($n = 67$), or a health education control group ($n = 67$) for a weekly intervention of 60 min for 40 weeks. The music program consisted of playing a percussion instrument, such as a conga, in weekly 60-min sessions for 40 weeks. Between 11 and 19 students participated in each class at the community center, and one or two professional music teachers were assigned to each session. At the beginning of the session, the instructor presented the score and demonstrated the percussion score several times to the participants. Participants were taught to memorize the rhythm and sequence of percussion blows and learn to read music. They then played the score until they were able to make minimal or no mistakes. In the later sessions, participants improvised individually and played together in two groups. At the completion of each session, the music teacher gave the participants homework assignments to perform the musical scores they had learned in that session. Over 40 weeks, the dance group's memory test scores improved compared to the control group, but the music group's did not. Compared to the control group, both the dance and music groups improved on tests of general cognitive function. A long-term cognitive leisure activity program that included dancing and playing musical instruments resulted in improved memory and general cognitive function compared to a health education program for older adults with mild cognitive impairment.

In the Guidelines for the Treatment of Dementia Diseases 2017 [11], music has been shown to have an effect on behavioral and psychological symptoms of dementia, while some studies, as mentioned above, show that musical performance can improve core symptoms of dementia. As more such studies are conducted in future, it is possible that evidence will be established that musical performance can improve core symptoms. However, many intervention studies have used percussion instruments, which are easy to play because the musical experience of the subjects is diverse. Therefore, it is desirable to develop new electronic instruments that do not depend on the musical experience of the subjects.

4 Dementia and Physical Exercises with Music

Many observational studies have shown that exercise is associated with a lower incidence of dementia and Alzheimer-type dementia. Regular exercise has been shown in many observational studies to prevent dementia in the elderly or reduce cognitive decline in the elderly and is actively recommended [11]. Furthermore, it has been shown that combining exercise with cognitive function training can be even more effective [12–14]. Therefore, the Mihama-Kiho Project conducted in the towns of Mihama and Kiho in Mie Prefecture, Japan investigated whether an intervention combining exercise and music in healthy elderly people is even more effective than exercise alone [15, 16]. Nationally, the aging rate in both towns has progressed to the point at which it is roughly equivalent to the average for Japan in the last 20 years. Therefore, the results of the intervention in both towns may be applicable to Japan for 20 years.

The target population comprised 207 healthy elderly people living in the area. The 166 people who wished to participate in the exercise class were divided into two groups: a music exercise group and an exercise group. Forty-one participants were assigned to the control group, without any intervention. The exercise group participants exercised once a week for one hour each for one year under the guidance of a professional instructor.

As an intervention, physical exercises with musical accompaniment were used as interventions. For the physical exercises-alone group, the exercises were the same as that of the musical exercises group, but, instead of musical accompaniment, only a drum beat was used. Neuropsychological examinations and brain MRI examinations were conducted before and after the intervention for the musical physical exercises and physical exercises groups, and twice at one-year intervals for the control group.

The following neuropsychological tests were administered: Mini-Mental State Examination (MMSE), Raven Chromatic Matrices Test (RCPM), Immediate/Delayed Replay of Logical Memory (LM-I/II), Word Recall (animal names, word starters), Trail-Making Test (TMT) A/B, and cube copying.

T1-weighted images at 1.5T were acquired as brain MRI scans. Brain morphometry was carried out using the SPM12.

In the neuropsychological testing, we conducted neuropsychological tests on 51 subjects in the music and exercises group, 61 subjects in the exercises group, and 32 subjects in the control group. In the neuropsychological test, the visuospatial cognitive function of the musical exercises group was significantly improved compared to that of the control group. Brain morphometry showed that the volume of the frontal lobe and auditory cortex in the musical exercises group, and the frontal lobe in the exercises group increased after the intervention compared to before the intervention (FWE $p < 0.05$). The volumes of the frontal lobe and hippocampus of the musical exercise group and the exercises group were maintained and increased compared to the control group. Although aerobic exercise has been shown to increase the volume of the hippocampus [17], the volumes of the frontal lobe and the hippocampus were found to have been maintained and increased, respectively, with musical accompaniment. Since the frontal lobe volume was maintained and increased in the order of musical exercise group > exercises group > control group, it is clear that musical exercises maintain and increase the frontal lobe volume as well as improve the cognitive function of healthy elderly people.

Satoh et al. [18] also showed that musical exercises maintained activities of daily living in patients with dementia. Eighty-five dementia patients (MMSE 15–26 points) were randomly divided into musical exercises (ExM) (43 patients) and brain training (BT) (42 patients) groups. ExMs were given musical exercises, and BTs were given portable games and drills, for 40 min once a week for 6 months. Neuropsychological tests were carried out prior to and after the study to examine cognitive function and activities of daily living.

The following neuropsychological tests were administered: MMSE, RCPM, LM-I/-II, Word Recall (animal names, initial sounds), TMT-A/B, and cube copying. Behavioral Pathology in Alzheimer's Disease (Behave-AD) and Functional Independence Measure (FIM) were used to assess daily living.

Sixty-two subjects, excluding 23 who dropped out, were analyzed. There were no differences in age, education, or starting MMSE scores between the two groups. A between-group comparison of the amount of change showed a significant improvement in ExM cube copying ($p = 0.009$), and a trend toward improvement in TMT-A ($p = 0.070$) and FIM ($p = 0.066$). The FIM did not change in the ExM group ($p = 0.385$), but worsened significantly in the BT group ($p = 0.048$).

5 Predicting the Effects of Non-pharmacological Therapies from Cognitive Function and Brain Volume in Dementia Patients

Previous studies have shown the efficacy of aerobic exercise in preventing and slowing the progression of dementia. In addition, exercise combined with cognitive training has been shown to have a greater effect on cognitive function than exercise alone. We have shown that exercises set to music (musical exercises: ExM) has a greater effect on visuospatial cognition in healthy elderly people than exercises alone, resulting in extensive morphological changes[15, 16]. Musical exercises was also shown to prevent deterioration in activities of daily living in patients with mild to moderate dementia better than brain training using portable game consoles and drills (Brain Training: BT) [18].

Few studies have focused on predictors of the effectiveness of non-pharmacological therapies. Identifying predictors may allow for more targeted non-pharmacological therapies to be offered, taking into account factors that may weaken the impact of interventions. In fact, previous studies [19, 20] have shown that clinical information, such as patient background and severity of illness, can influence the effectiveness of non-pharmacological therapies. However, the neuropsychological factors influencing the effects of non-pharmacological therapies and their neural basis remain unknown. Furthermore, previous studies have not compared different non-pharmacological therapies in randomized controlled trials (RCTs).

The Mihama-Kiho Project sought to assess pre-intervention cognitive function and brain volume in order to identify predictors of the effects of non-pharmacological therapies on patients with mild to moderate dementia [21].

The subjects were patients with mild to moderate dementia living in the towns of Mihama or Kiho in Mie Prefecture, Japan and who were using nursing care services. The subjects were being treated with anti-dementia drugs, which did not change during the intervention period. The subjects were randomly divided into two groups to receive either a musical exercise or a brain training intervention (once a week, 40 min, for 6 months). In the musical exercises, the subjects sat on a chair and bent and stretched their arms and hips, or stepped on their feet to the rhythm of pop music with varying tempo. In the brain training session, the subjects performed simple calculations, mazes, and image error problems using portable game consoles and drills. Neuropsychological tests including intellectual functioning (MMSE, RCPM), memory (LM-I/II), frontal lobe functioning (word recall [animal names, word starters], TMT), visuospatial cognition (cube copying), and activities of daily living (FIM). In the brain MRI examination, T1-weighted images at 1.5T were taken. Brain morphometry (voxel-based morphometry) was performed using MATLAB and Statistical Parametric Mapping 12.

Forty-six patients with dementia (25 with musical exercises and 21 with brain training) who participated in at least 75% of the musical exercises or brain training interventions and who underwent neuropsychological testing and MRI imaging both before and after the interventions were included in the analysis. Following a previous study, we divided the patients into two groups based on their MMSE scores after 6 months: 18 patients in the improvement group (11 in musical exercises, 7 in brain training, 17 in AD, 1 in VaD, MMSE $+3.8 \pm 1.6$) and 28 patients in the non-improvement group (14 in musical exercises, 14 in brain training, 22 in AD, 6 in VaD, MMSE -1.52 ± 1. There was no significant difference in MMSE scores between the improvement and non-improvement groups before the intervention. Cognitive function in the non-improvement group prior to the intervention was significantly worse than that of the improvement group in memory in the musical exercise, and in intellectual function and cognitive items of daily life activities in the brain training. The brain volume of the non-improvement group before the intervention was smaller in the anterior cingulate cortex in the musical exercise group and in the left middle frontal gyrus in the brain training group than in the improvement group.

The differences in cognitive function shown by the results may reflect differences in the cognitive resources needed to implement each intervention. Participants needed memory to remember movement patterns for the musical exercises, and general intellectual function to perform calculations, mazes, and search for errors in images for brain training. In addition, previous studies [22] have shown that the brain regions that the results indicate play an important role in the execution of each intervention. While the study showed that patients with mild to moderate dementia with more advanced cognitive decline and extensive cortical atrophy are less likely to show improvement in cognitive function after non-pharmacological treatment, the assessment of cognitive function and brain volume prior to intervention may help select the type of non-pharmacological treatment that would be effective.

6 Towards More Effective Non-pharmacological Therapies for Dementia

Given the worldwide increase in dementia and the lack of a cure, non-pharmacological therapies are very important. In the past few years, many non-pharmacological therapies, including exercise, have been proven effective. Just as medication should be administered based on patient information, non-pharmacological therapies should not only be administered in the dark, but should also be administered with patient information in mind based on applied informatics. Thus, non-pharmacological treatment for dementia may become more effective.

Table 1. Content of non-pharmacological treatment (modified from Guidelines for the Treatment of Dementia Diseases, 2017, supervised by the Japanese Society of Neurology).

Non-pharmacological treatment	Overview
Cognitive function training	Focuses on specific areas of cognitive functioning, such as memory, attention, and problem solving, and uses paper and computer-based tasks that are tailored to the individual's level of functioning.
Exercise therapy	A wide variety of programs exist. Programs ranging from 20 to 75 min, twice a week to every day, have been reported. The exercises are classified into aerobic exercise, muscle strengthening training, and balance training, and the programs are often composed of combinations of these multiple exercises.
Reminiscence therapy	It focuses on the past life history of the elderly and aims to support the individual through receptive, empathetic, and supportive listening to the life history.
Music Therapy	A wide variety of programs exist. Programs have been reported to be held one to five times a week for 10 to 60 min. There are methods such as listening to music, singing, playing percussion instruments, etc., and rhythmic exercises, and these are often combined to form a program.

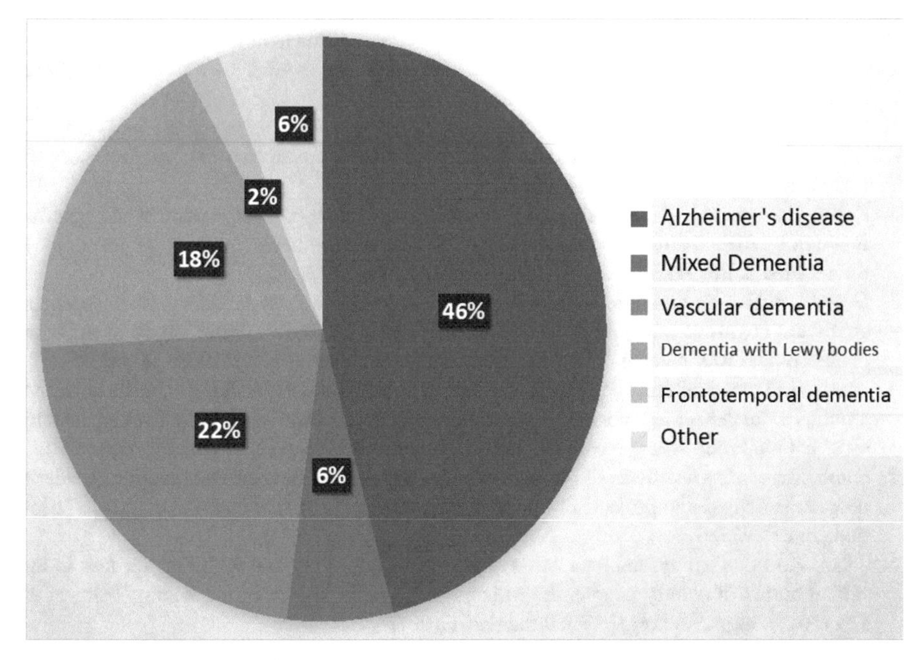

Fig. 1. Dementia rates (Akatsu et al., 2002).

References

1. Prince, M., Bryce, R., Albanese, E., Wimo, A., Ribeiro, W., Ferri, C.P.: The global prevalence of dementia: a systematic review and metaanalysis. Alzheimers Dement **9**(1), 63–75 e2 (2013). https://doi.org/10.1016/j.jalz.2012.11.007
2. Nakajima, K., Shimohama, S., Tomimoto, H., Mimura, M., Arai, T.: Ninchisho Handobukku (Handbook of dementia), 2nd ed. 医学書院Igakushoin 2020, pp. xxvii, 916p (2020).
3. Akatsu, H., et al.: Subtype analysis of neuropathologically diagnosed patients in a Japanese geriatric hospital. J. Neurol. Sci. **196**(1–2), 63–69 (2002). https://doi.org/10.1016/s0022-510x(02)00028-x
4. Mecocci, P., Boccardi, V.: The impact of aging in dementia: It is time to refocus attention on the main risk factor of dementia. Ageing Res. Rev. **65**, 101210 (2021). https://doi.org/10.1016/j.arr.2020.101210
5. Zhang, Y., et al.: Does music therapy enhance behavioral and cognitive function in elderly dementia patients? A systematic review and meta-analysis. Ageing Res. Rev. **35**, 1–11 (2017). https://doi.org/10.1016/j.arr.2016.12.003
6. Cowles, A., et al.: Musical skill in dementia: a violinist presumed to have Alzheimer's disease learns to play a new song. Neurocase **9**(6), 493–503 (2003). https://doi.org/10.1076/neur.9.6.493.29378
7. Cho, H., et al.: Postmorbid learning of saxophone playing in a patient with frontotemporal dementia. Neurocase **21**(6), 767–772 (2015). https://doi.org/10.1080/13554794.2014.992915
8. Verghese, J., et al.: Leisure activities and the risk of dementia in the elderly. N. Engl. J. Med. **348**(25), 2508–2516 (2003). https://doi.org/10.1056/NEJMoa022252
9. Balbag, M.A., Pedersen, N.L., Gatz, M.: Playing a musical instrument as a protective factor against dementia and cognitive impairment: a population-based twin study. Int. J. Alzheimers Dis. **2014**, 836748 (2014). https://doi.org/10.1155/2014/836748

10. Doi, T., et al.: Effects of cognitive leisure activity on cognition in mild cognitive impairment: results of a randomized controlled trial. J. Am. Med. Dir. Assoc. **18**(8), 686–691 (2017). https://doi.org/10.1016/j.jamda.2017.02.013
11. Committee for the Development of Guidelines for the Treatment of Dementia Diseases and T. J. S. o. Neurology, Ninchishogaidorain (Guidelines for the Treatment of Dementia Diseases). Igakushoin, 2017, p. xxii, p. 362 (2017)
12. Fabre, C., Chamari, K., Mucci, P., Masse-Biron, J., Prefaut, C.: Improvement of cognitive function by mental and/or individualized aerobic training in healthy elderly subjects. Int. J. Sports Med. **23**(6), 415–421 (2002). https://doi.org/10.1055/s-2002-33735
13. Oswald, W.D., Gunzelmann, T., Rupprecht, R., Hagen, B.: Differential effects of single versus combined cognitive and physical training with older adults: the SimA study in a 5-year perspective. Eur. J. Ageing **3**(4), 179 (2006). https://doi.org/10.1007/s10433-006-0035-z
14. Shatil, E.: Does combined cognitive training and physical activity training enhance cognitive abilities more than either alone? A four-condition randomized controlled trial among healthy older adults. Front. Aging Neurosci. **5**, 8 (2013). https://doi.org/10.3389/fnagi.2013.00008
15. Satoh, M., et al.: The effects of physical exercise with music on cognitive function of elderly people: Mihama-Kiho project. PLoS ONE **9**(4), e95230 (2014). https://doi.org/10.1371/journal.pone.0095230
16. Tabei, K.I., et al.: Physical exercise with music reduces gray and white matter loss in the frontal cortex of elderly people: the Mihama-Kiho scan project. Front. Aging Neurosci. **9**, 174 (2017). https://doi.org/10.3389/fnagi.2017.00174
17. Erickson, K.I., et al.: Exercise training increases size of hippocampus and improves memory. Proc. Natl. Acad. Sci. U. S. A. **108**(7), 3017–3022 (2011). https://doi.org/10.1073/pnas.1015950108
18. Satoh, M., et al.: Physical exercise with music maintains activities of daily living in patients with dementia: Mihama-Kiho project part 21. J. Alzheimers Dis. **57**(1), 85–96 (2017). https://doi.org/10.3233/JAD-161217
19. Sarkamo, T., Laitinen, S., Numminen, A., Kurki, M., Johnson, J.K., Rantanen, P.: Clinical and demographic factors associated with the cognitive and emotional efficacy of regular musical activities in dementia. J. Alzheimers Dis. **49**(3), 767–781 (2016). https://doi.org/10.3233/JAD-150453
20. Hsu, T.J., Tsai, H.T., Hwang, A.C., Chen, L.Y., Chen, L.K.: Predictors of non-pharmacological intervention effect on cognitive function and behavioral and psychological symptoms of older people with dementia. Geriatr. Gerontol. Int. **17**(Suppl 1), 28–35 (2017). https://doi.org/10.1111/ggi.13037
21. Tabei, K.I., et al.: Cognitive function and brain atrophy predict non-pharmacological efficacy in dementia: the Mihama-Kiho scan project2. Front. Aging Neurosci. **10**, 87 (2018). https://doi.org/10.3389/fnagi.2018.00087
22. Maurer, S., et al.: Non-invasive mapping of calculation function by repetitive navigated transcranial magnetic stimulation. Brain Struct. Funct. **221**(8), 3927–3947 (2015). https://doi.org/10.1007/s00429-015-1136-2

Research Methods to Build a Reference Model for Designing Addiction-Aware Video Game

Flourensia Sapty Rahayu[1,2](✉), Lukito Edi Nugroho[1,2], and Ridi Ferdiana[1,2]

[1] Departemen Teknik Elektro & Teknologi Informasi, Universitas Gadjah Mada, Jl. Grafika No. 2, Sleman, DIY, Indonesia
sapty.rahayu@uajy.ac.id

[2] Prodi Sistem Informasi, Universitas Atma Jaya Yogyakarta, Jl. Babarsari 43, Sleman, DIY 55281, Indonesia

Abstract. Previous research have studied the various risk factors that can lead to video game addiction. One environmental factor that contributes to the development of addiction is video game design. Based on this fact, we argue that the design aspects can also be potentially used to minimize the compulsive effect of video games. This effort may be done by conducting a study to develop principles, guidelines, or models that can be used by designers to design less addictive-video games. The selection of proper methodology and methods to perform the research is significant to do. This present study aims to select the proper methods for the next research on developing a reference model for designing addiction-aware video games. The results show four phases of the proposed research and the selected methods for each phase with its justifications. Some HCI methods selected include observation, survey, diary study, interview, focus group, and experiment.

Keywords: research method · human-computer interaction (HCI) · video game addiction · Iiternet gaming disorder · reference model

1 Introduction

A video game is one of the information technology applications that can be used by nearly all ages. Some motivations people play games include getting entertainment and leisure [4], to escape from problems, and to reduce stress. One factor causing video games to become very popular is the human need for feeling comfortable and happy. Video games have some benefits and disadvantages. One of the disadvantages is the onset of addiction. Video game addiction is one form of behavioral addiction characterized by salience, mood modification, tolerance, withdrawal, conflict, and relapse [12]. Like many other addictions, game addiction can lead to more negative outcomes, such as a decreasing academic performance, deterioration of interpersonal relationships [13, 14, 25], and may contribute to behavior problems, such as aggression and delinquency [17]. Video game addiction can be experienced by all age groups. A literature study by Darvesh et al. [8] found the prevalence of video game addiction in the general population from

T. Matsuo et al. (Eds.): AIMD 2019, LNNS 677, pp. 111–120, 2023.
https://doi.org/10.1007/978-3-031-30769-0_11

160 studies as follows: children and adolescents (aged 0–19 years) have a prevalence rate 0,26–38%, while adults (aged 18-older) have prevalence rate 0,21–55,77%.

Now, video game addiction has become a global issue related to public health around the world. Internet Gaming Disorder (IGD) has been mentioned in section III of the fifth edition of the Diagnostic and Statistical Manual of Mental Disorders (DSM-5) as a condition that needs further examination before considered a formal disorder. Although IGD has not been formally recognized yet as a disorder, there have been some strategies to intervene in the IGD. The intervention strategies commonly used to treat internet addiction (including video game addiction) are adopted from other substance addiction strategies. These strategies are used with the understanding that all addictions share common characteristics. Some strategies include psychotherapy, pharmacology, or a combination of both [22]. Some countries, such as South Korea and China have made regulations to manage the problematic use of video games in their countries [20].

While not all individuals develop an addiction when playing video games, gamers who become addicts are likely to have biological and genetic factors, environmental influences, and developmental elements that contribute to their addictive behavior [9]. One environmental influence that contributes to the onset of addiction is the technology design. Some researchers argue that technology design is responsible for the development of addiction ([5, 7, 9, 10, 15]). Some design principles are purposely used to create a compulsive effect on users ([6, 23]). Those principles or features are also implemented in video game design. Based on the fact that addiction is a purposely created behavior, we argue that the design aspects can also be potentially used to minimize the compulsive effect of video games. This effort can be done by exploring principles, guidelines, or model to design an addiction-aware video game. These principles, guidelines, or models can be used by the designer to design a video game with the consideration to minimize the addiction effect.

Previous studies on the design aspects of video games contributing to the onset of addiction are scarce. A preliminary study by Alrobai and Dogan [3] investigated the concept of digital addiction to build a requirements engineering framework for addiction-aware software. However, that study was still in the conceptual planning stage. We cannot find further research nor the realization of the concept. Another study investigated the development of a reference model to build a reference model designing interactive online platforms to combat digital addiction [2]. However, the reference model proposed [2] was not designed to guide the development of software that can cause addiction itself. The two studies above were done in the broader context of digital addiction. However, we did not find similar studies, particularly in the video game domain. Thus, we propose a research to develop a reference model for designing an addiction-aware video game. Since the study of design principles, guidelines, or models lies in the human-computer interaction (HCI) field, the proper HCI methods to create design principles, guidelines, or models are significant to be chosen. This paper is a part of the overall proposed research for designing an addiction-aware video game. This paper aims to explore the research methods in HCI and find the proper methods for developing a reference model that can be used to design an ad-diction-aware video game.

2 Literature Review

Some common research methods in HCI include observations, surveys, diary studies, interviews, focus groups, controlled experiments, case studies, and ethnography [21]. The observation method is a systematic method to observe and record description notes, individuals' or groups' behavior analysis, and interpretations [24]. There are two types of observations: (1) participant observation, and (2) structured observation. Participant observation is a qualitative approach to discover people's action meanings. This method needs full participation from the researcher in the subjects' lives and activities and thus becomes a member of their community. The researchers can share their experiences not only by observing what is happening but also by feeling it [11]. Structured observation is a quantitative approach to investigate the frequencies of people's actions. It requires a pre-determined structure, e.g. a list of the behavioral aspects and their contextual factors, to quantify the behavior.

The survey is a method to gather responses from an individual with a set of questions that are well-defined and well-written. Surveys are often used for describing populations, explaining behaviors, and exploring uncharted waters [21]. The survey can get a large number of responses and capture the problem's "big picture" quickly. But there are some drawbacks to the surveys. Researchers can not get "deep" detailed data from the surveys, and sometimes the results of the survey can be biased. The number of participants contributing to the surveys needs to be determined properly with a particular sampling technique. Two sampling techniques can be used: (1) probabilistic sampling (random sampling), and (2) non-probabilistic sampling. Probabilistic sampling aims at providing population estimates [21]. This technique is suitable to use when a population can be well defined. However, when there is not a well-defined population, we can not use probabilistic sampling. The alternative is to use a non-probabilistic sampling. There are some types of non-probabilistic sampling [4]: (1) convenience sampling, (2) judgmental sampling, (3) quota sampling, (4) snowball sampling, and (5) saturation sampling.

A diary is a document that contained regular recordings of an individual's life events, recording at the time that those events occur [1]. Diary is a suitable method to use if researchers want to record information that is fluid and changes over time. Diary is also suitable to use for understanding behaviors or situations that are not well-understood. Diary is an HCI research method filling the gaps between observation in real settings, observation in a lab, and surveys [18]. It is suitable to be used in a situation where it is not feasible to either bring users into a fixed setting or visit the users in their natural setting. Diaries are also good to study cross multiple technologies, multiple environments, and location usage patterns [16]. Since the nature of the data in a diary is represented in a qualitative nature format, content analysis is used to analyze the data.

Interview and focus group are methods that are used to explore deep responses from participants through direct conversations [21]. The interview is a direct discussion with the individual, whether focus groups is a direct discussion involving multiple users at one time. According to the level of formality and structure, interviews can be categorized as: structured interviews, semi-structured interviews, and unstructured or indepth interviews. A predetermined and 'standardized' set of questions is used in structured interviews. Semi-structured interviews do not use predefined questions, but only use the list of themes to be covered, so the questions may vary from interview to interview. In unstructured interviews or 'non-directive' interviews, there is also no predetermined list of questions, so the interviewee can talk freely about behavior, events, and beliefs related to the topic area.

The experiment method is used to identify the causal relationships between two or more factors. In a classic experiment, two groups of participants are given a random assignment. The first group is called the experimental group. A planned intervention or manipulation is given to this group. Whether in the second group, the control group, there is no such intervention given. This method measures the dependent variable before and after the manipulation of the independent variable for both groups. The experimental research is considered effective to generalize the findings to larger populations. However, it has some limitations, such as it needs tight control of factors that may influence the dependent variables, it needs well-defined and testable hypotheses, and the lab-based experiments' results may not represent a users' typical interaction behavior. There are three types of experiment methods, they are (1) true experiments, (2) quasi-experiments, and (3) non-experiments. A true experiment is a study involving multiple groups or conditions and the participants are randomly assigned to each condition. A quasi-experiment is an experiment where there are multiple groups or conditions, but there are no random assignments on the participants to different conditions. Whereas a non-experiment is an experiment where there is only one observation group or only one condition involved.

Table 1. The Benefits and Drawbacks of Research Methods

Research Methods	Benefits	Drawbacks
Observation	• Observes and records actual behavior • Have high external validity • Flexible	• Ethical concern about spying on people • It can not examine the causes of the problem
Survey	• Easy to collect data from a large number of people, at a relatively low cost • It can be very useful for getting an overview of a user population • Do not require advanced tools for development	• Not very good at getting "deep," detailed data • It can sometimes lead to biased data
Diary study	• **Record experiences in a natural environment** • **More likely to capture influential external factors** • **Collect observations in longer durations** • **More time for in-depth consideration and opportunities for creativity**	• The quality of the results depends on the participants • Time-consuming • **Researchers don't get to observe participants**

(*continued*)

Table 1. (*continued*)

Research Methods	Benefits	Drawbacks
Interview and focus group	• 100% response rate • Flexible (groups or individual interviews) • Efficient to administer with groups	• Time-consuming (with individual interviews) • Risk of interviewer bias
Experiment	• Allow the researcher to control the situation • Permit the researcher to identify cause and effect	• The situation is artificial and results may not generalize well to the real world • Sometimes difficult to avoid experimenter effects
Case study	• Flexibility • Capturing reality • Provide a rich and deep understanding of the problem	• The result may not be generally applicable • Selecting cases are often difficult • Difficult to provide external validity
Ethnography	• Provide a comprehensive perspective • Observes behavior in their natural environments	• Time-consuming and required a well-trained researcher • Dependent on the researcher's observations and interpretations • May lack transferability

Our proposed research aims to develop principles and the reference model for designing an addiction-aware video game. This research will involve gamers as research participants. Overall, the proposed study will start with the exploration of user experience and game features that potentially lead to addiction. Next, the development of principles and reference model will be conducted based on the results of the exploration activities. The last stage will be the validation of the results. Based on the benefits and drawbacks of HCI research methods as seen in Table 1, several methods may be suitable to use to conduct the data collection or result validation. Here are some potentially used HCI methods for the proposed study:

- *Observation.* This method may be suitable to use for exploring user experiences if it is possible to observe participant activities in a real environment.
- *Survey.* This method is may be suitable to use in the early phase for selecting participants that match the characteristics of game addicts.
- *Diary study.* This method may be used to explore the user experience during gaming. A Diary study may be suitable to use if it is not possible to do an observation in a laboratory nor the real environment.
- *Interview and focus group.* These methods may be used to gain a deeper understanding of the problem or to confirm results from participants or experts.

- *Experiment*. An experiment may be used to validate the results of the proposed study empirically.
- *Ethnography*. Ethnography may be used to gain a deeper understanding of user experiences in gaming if it is possible to do in the real environment.

The case study method may not be suitable to use for the proposed study since the result of the case study project may not be generally applicable. Whether the pro-posed study aims to produce a model that can be applied generally.

3 The Chosen HCI Methods to Develop Reference Model in Designing Addiction-aware Video Game

In the next study, we plan to build a reference model that can be used to guide designers in developing an addiction-aware video game. To achieve the objective, the proper methods need to be chosen. Our study uses a mixed-method that combines qualitative and quantitative methods. The early phase of the study will focus on the exploration of user behavior and game artifacts that potentially lead to addiction. Since the activities will involve human perception, the qualitative approach is considered suitable to be used. The qualitative approach will be used mostly in the early phase of the study. The late phase of the study, particularly the validation stage, will apply the quantitative approach. The quantitative method will be used to do the empirical validation. The overall stages of the research will be divided into four phases, namely, the orientation, exploration, investigation, and confirmation phases (Fig. 1).

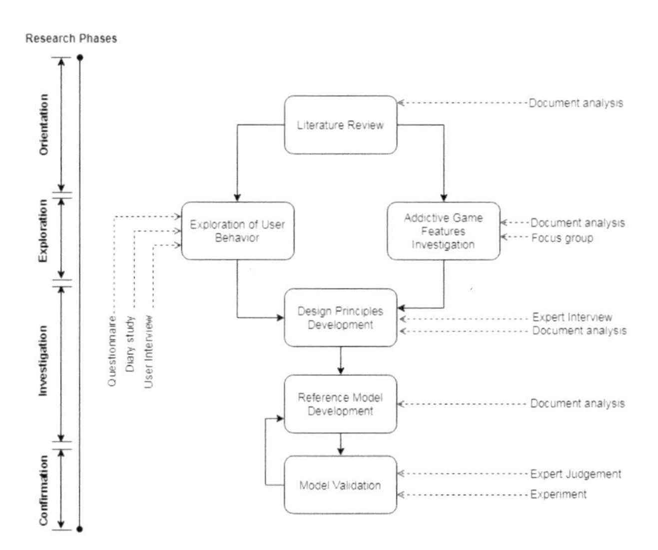

Fig. 1. Research Phases

The orientation phase focuses on the topic formulation, contextualization, and limitation of the problem. We did a literature review to get insight into the problem domain and to formulate the topics, The literature review was done by selecting the articles from reputable journals and conferences indexed in Scopus. In this stage, we used the document analysis method to collect data and the content analysis method to analyze data. The literature searched and analyzed covered broad topics of digital addiction, including the use of information technology to combat digital addiction. The research gaps found from this literature review is used as a basis for proposing the research.

The second phase, the exploratory phase, aims to explore some issues related to video game addiction. In this exploratory phase, we will explore the user's behavior and the features of video games that potentially be the trigger of addiction. This phase consists of two steps. The first step aims to explore user behavior or user experience when using video games. By knowing the user behavior or user experience when using the game, researchers are expected to get information about the factors that support the emergence of addiction. Some data collection methods will be used, namely, questionnaire, diary study, and interview. Questionnaires are used to select the research participants who meet the criteria of game addiction using the Indonesian Online Game Addiction Questionnaire [19]. This questionnaire will be presented online. The questionnaire is considered a suitable tool to achieve this goal. Participants will be selected using the judgmental sampling method. Judgmental sampling is the selection of participants based on the knowledge and judgment of researchers about the most representative subjects in the population [4]. The participants must meet the following criteria: (1) young adults (aged 18–24 years), (2) the duration of game playing is at least three hours per day and the frequency of game playing is at least four days per week. Some of the participants that meet the video addiction criteria will be asked to join the focus group discussion in the next step.

After the subjects are selected, the data collection will be performed. One of the data collection methods that will be used is a diary study. Diary study can bridge the gap in human-computer interaction research between observations in their natural state, observations in the laboratory, and surveys [20]. This method is considered suitable for this study since we need to observe and get the experience of the participants when playing video games in their natural setting for a few months. This activity will not be done in the laboratory because it may influence the user experience, nor it will not be done with true observation due to the time and cost limitation of the research. In this study, research subjects will be asked to record their daily activities in playing video games. Mobile applications will be used as a diary. Another method to collect data in this phase is the interview. The interview method is a direct discussion with specific goals between two or more people to get detailed information [16]. Interviews can be used for initial exploration, exploring needs, and evaluating [3]. This method is considered suitable to use for gaining deeper information from the participants personally. In this phase, interviews with participants were used to confirm and explore deeper information based on the results obtained from the diary study.

The second step in the exploration phase is the exploration of potentially addictive game features. The knowledge of potentially addictive game features will be used as a basis to formulate the design principles. At this stage, two techniques will be used for data collection, namely document analysis, and focus groups. Document analysis is a systematic procedure for reviewing and evaluating documents, whether in printed or electronic form [23]. Document analysis is also often used as a tool in triangulation to reduce the researcher's bias. In this research, document analysis will be performed by analyzing data from journal articles, conferences, online articles, and web forums to explore information about game features. Another technique to be used is the focus group. A Focus group is an interactive discussion that is regulated by a moderator that maintains and controls the discussion focus [16]. This method is appropriate to get responses from several research samples (generally between four and eight, up to 12 participants) in one activity. In this study, the focus group will be used to explore more information about the addictive game features from research participants. This activity will be done by an online meeting. The participants are selected from the questionnaire respondents that meet the video game addiction criteria. A focus group is selected instead of an interview since group discussion may generate a more comprehensive understanding of the game features identified. The findings from this focus group are used to complement, refine, or validate the findings obtained from the document analysis.

The results of the exploration phase will be used as an input for the investigation phase. In the investigation phase, there will be two activities performed, namely the development of design principles and the construction of reference models. The first activity aims to investigate the design principles that will be used to develop a reference model. Data collection techniques used in this stage are document analysis and interviews with the experts. Document analysis will be done by analyzing literature related to the design principles, from journal articles, conferences, and online articles. Document analysis will also be performed by studying theories related to the design principles from psychology, social science, and computer science. Interviews with experts will be conducted to obtain more information and to get confirmation about the findings obtained in document analysis. The expected participating experts are those who have competence in the field of video game design or human-computer interaction.

The second step in the investigation phase is the development of reference models. Based on the design principles produced in the previous stage, a reference model will be developed. To develop a reference model, the researchers will also use the knowledge gained from document analysis. As in the previous stage, document analysis will be done by analyzing the literature related to the construction of reference models, from journal articles, conferences, and online articles.

The last phase, a confirmation phase, is needed to ensure that the model produced is valid. In this phase, the reference model will be validated. Model validation will be performed with two techniques, namely, expert judgment and experiment. Expert Judgment is a technique for making an assessment where the assessment is made based on a set of criteria and/or specific expertise that has been obtained by someone in a particular field of knowledge or discipline [24]. Three to five experts in the field of video game design or human-computer interaction will be involved to review the findings and model and check their validity. To strengthen the validity of the findings, empirical model validity testing will also be performed by experimenting. This experiment method is considered suitable to use for proving that the principles and a model produced from this study are empirically validated. This study will use the true experiment method in which there are two groups and two conditions in-volved and participants are assigned randomly to each condition. To get the participants, the purposive/judgmental sampling method will be used. Participants will be divided into two groups, namely, the experimental group and the control group. An experimental group is a group that is given special treatment. The special treatment in this experiment is to assign the experimental group to use a game prototype that has implemented the principles and models produced. The experiment will be performed with the following scenarios:

a) The researchers will develop two game prototypes. Prototype 1 will implement the proposed principles and models, while prototype 2 will not implement the proposed principles and models.
b) The participants will be divided into two groups. The first group (experimental group) will be asked to use prototype 1 for 3–6 months, while the second group (control group) will be asked to use prototype 2 for the same period.
c) Addiction measurements will be performed for the two groups by adapting the Indonesian Online Game Addiction Questionnaire measurement tool [22].
d) The measurement results for each group will be compared and analyzed. Analysis of the measurement results will use the inferential statistics method.
e) Steps a - d will be repeated by exchanging game prototypes that must be used for both groups.

4 Conclusion

This study aims to analyze the proper methods that can be used to conduct the next proposed research on developing a reference model for designing addiction-aware video games. The proposed development of a reference model has four phases: orientation, exploration, investigation, and orientation. Some commonly used methods in HCI are selected to performs the planned activities in each phase. The methods considered suitable to use in this study include observation, survey, diary study, interview, focus group, and experiment. The selection of the method is based on the goal of each activity and the strengths and weaknesses of each method. Some HCI methods that will be used in the exploration phase are: survey with the questionnaire, diary study, user interview, and focus group. The investigation phase will employ an interview method to get information from the experts. In the orientation phase, the experiment method will be used other than expert judgment, to validate the principles and model produced empirically. Since this paper is only one part of the overall research to develop a reference model for designing an addiction-aware video game, the effectiveness of the methods selected has not been proven yet.

References

1. Alaszewski, A.: Using Diaries for Social Research. Sage Publications Ltd, London (2006)
2. Alrobai, A., et al.: COPE.er method: combating digital addiction via online peer support groups. Int. J. Environ. Res. Public Health **16**, 7 (2019). https://doi.org/10.3390/ijerph16071162
3. Alrobai, A., Dogan, H.: Requirements engineering for ADDICTion-Aware software (E-ADDICT). In: CEUR Workshop Proceedings, vol. 1138, pp. 46–52 (2014)

4. Alrobai, A.A.: Engineering Social Networks to Combat Digital Addiction: The Case of Online Peer Groups. Bournemouth University (2018)
5. Alter, A.: Irresistible: the rise of addictive technology and the business of keeping us hooked. Penguin Group (USA) LLC (2017)
6. Alutaybi, A., et al.: How can social networks design trigger fear of missing out? How can social networks design trigger fear of missing out? In: IEEE International Conference on Systems, Man, and Cybernetics (IEEE SMC 2019) (2019)
7. Berthon, P., et al.: Addictive de-vices: a public policy analysis of sources and solutions to digital addiction. J. Public Policy Mark. **38**(4), 451–468 (2019). https://doi.org/10.1177/0743915619859852
8. Darvesh, N., et al.: Exploring the prevalence of gaming disorder and Internet gaming disorder: a rapid scoping review. Syst. Rev. **9**, 1 (2020)
9. Doan, A.P., Strickland, B.: Hooked on Games: The Lure and Cost of Video Game and Internet Addiction. F.E.P. International, Inc. (2012)
10. Eyal, N.: Hooked: How to Build Habit-Forming Products. Penguin Group (USA) LLC, New York (2014)
11. Gill, J., Johnson, P.: Research Methods for Managers. Sage Publications Ltd, London (2002)
12. Griffiths, M.: Does internet and computer addiction exist? Some case study evidence. Cyberpsychol. Behav. **3**(2), 211–218 (2000). https://doi.org/10.1089/109493100316067
13. Griffiths, M.D., et al.: Online computer gaming: a comparison of adolescent and adult gamers. J. Adolesc. **27**, 87–96 (2004). https://doi.org/10.1016/j.adolescence.2003.10.007
14. Griffiths, M.D., et al.: Video game addiction: past, present and future. Curr. Psychiatry Rev. **8**(4), 308–318 (2012). https://doi.org/10.2174/157340012803520414
15. Harris, T.: How a Handful of Tech Companies Control Billions of Minds Every Day. Accessed 30 March 2020
16. Hayashi, E., Hong, J.I.: A diary study of password usage in daily life. In: Proceedings of the SIGCHI Conference on Human Factors in Computing Systems, pp. 2627–2630 (2011)
17. Holtz, P., Appel, M.: Internet use and video gaming predict problem behavior in early adolescence. J. Adolesc. **34**(1), 49–58 (2011). https://doi.org/10.1016/j.adolescence.2010.02.004
18. Hyldegård, J.: Using diaries in group based information behavior research- a methodological study. In: Proceedings of the Information Interaction in Context, pp. 153–161 (2006)
19. Jap, T., et al.: The development of Indonesian online game addiction questionnaire. PLoS ONE **8**(4), 4–8 (2013). https://doi.org/10.1371/journal.pone.0061098
20. Király, O., et al.: Policy responses to problematic video game use: s systematic review of current measures and future possibilities (2017). https://doi.org/10.1556/2006.6.2017.050
21. Lazar, J., et al.: Research Methods in Human-Computer Interaction. Morgan Kaufmann, Cambridge, MA (2017)
22. Mihajlov, M., Vejmelka, L.: Internet addiction: a review of the first twenty years. Psychiatr. Danub. **29**(3), 260–272 (2017). https://doi.org/10.24869/psyd.2017.260
23. Montag, C. et al.: Addictive features of social media/messenger platforms and freemium games against the background of psychological and economic theories. Int. J. Environ. Res. Public Health **16**(14), 2612 (2019). https://doi.org/10.3390/ijerph16142612
24. Saunders, M., et al.: Research Methods for Business Students. Pearson, Essex (2009)
25. Smyth, J.M.: Beyond self-selection in video game play: an massively multiplayer online role-playing game play. Cyberpsychol. Behav. **10**(5), 717–721 (2007). https://doi.org/10.1089/cpb.2007.9963

TrAcOn: A Traffic Accident Ontology for Identifying Accidents-Prone Areas in Senegal

Mouhamadou Gaye[1(✉)], Ibrahima Diop[1], Ana Bakhoum[2], and Papa Alioune Cissé[1]

[1] Computer Science and Engineering for Innovation Laboratory, Assane Seck University, Ziguinchor, Senegal
{m.gaye,ibrahima.diop,pa.c1}@univ-zig.sn
[2] Assane Seck University, Ziguinchor, Senegal
a.bakhoum4003@univ-zig.sn

Abstract. Traffic accidents are responsible each year for many deaths in the world. Most of them are caused by human fault (speeding, driving while intoxicated, dangerous overtaking, drowsiness, dangerous parking, etc.). Despite the progress noted with the arrival of smart cars and the semantic information representation through ontologies, underdeveloped countries have not yet integrated IT solutions into their transport management systems. Added to this, the absence of concepts that can describe certain types of transport vehicles in ontologies developed for transport systems. Our research objective is to propose an approach to identifying areas at high risk of accidents in less developed countries like Senegal. In this article, we build an ontology of traffic accidents, called TrAcOn (Traffic Accident Ontology) for a traffic accident description system in Senegal. The ontology allows among others to identify highly accidental areas.

Keywords: Traffic accident · Ontology · Descriptions · Formaling · Information representation

1 Introduction

In Senegal as in the world, traffic accidents are responsible each year for many deaths. According to Senegal Ministry of transportation, road insecurity costs at least 163 billion CFA francs per year, or 2% of GDP (Gross Domestic Product). The WAEMU (West African Monetary Economic Union), considering the United Nations resolution on road safety, has enacted the Directive 14-2009CM-UEMOA[1] whose objective is to allow member states to have an information system of traffic accident. In fact, most of underdeveloped countries, especially West Africa countries, have not yet integrated IT solutions into their transport management systems.

Our goal is to offer a computer system for traffic accidents management in Senegal, which can be adapted in other African countries. The system we propose is based on a domain ontology of traffic accidents that reuses existing ontologies in its conception.

[1] http://www.uemoa.int/sites/default/files/bibliotheque/directive_14_2009_cm_uemoa.pdf.

T. Matsuo et al. (Eds.): AIMD 2019, LNNS 677, pp. 121–131, 2023.
https://doi.org/10.1007/978-3-031-30769-0_12

However, the particular context of road traffic in Senegal characterized by an obsolete car fleet, impassable roads, a lack of traffic signs, etc. involves the introduction of ontological concepts such as: "Car-Rapide", "Ndiaga-Ndiaye", "Clando", "Djakarta", "Charrette", "Pousse-Pousse" which describe specific transport vehicles.

In this article, we construct a domain ontology called Traffic Accident Ontology (TrAcOn) which takes into account all the factors implied in a traffic accident in Senegal and that will be integrated in the road accident management system.

The rest of the article is organized as follows. In Sect. 2, we present the context and ontological needs. In Sect. 3, we present the ontology building by starting with the delimitation of the domain and the scope. Section 4 is devoted to formalization where the ontology is represented by a model. We end by a summary and outlook.

2 Context and Ontological Needs

2.1 Context

Traffic accidents are one of the leading causes of death in the world, especially among people between the age of 5 and 29 according to the World Health Organization report on road safety [1]. Still in this report, the number of deaths caused by traffic accidents reached 1.35 million in 2016 and no low- income country has seen a decline in the number of road traffic fatalities. In Africa, the death rate from traffic accidents is the highest in the world (26.6/100,000 inhabitants). In fact, most of underdeveloped countries, especially West Africa countries, have not yet integrated IT solutions into their transport management systems. This is generally due to the lack of financial resources. Indeed, most African countries are struggling to allocate up to 8% of their GDP to the transport sector according to the 2010 African transport review report of the United Nations Economic Commission for Africa[2]. This situation is so different from that of developed countries which have revolutionized their transport systems in recent years with the advent of Intelligent Transport Systems (ITS). In developed countries, ITS are present in roads management and motorways, transit management systems, incident management systems, etc. They involve Semantic web and ontologies as a solution to the lack of semantics in the information representation (accidents descriptions) and inference.

Ontologies allow autonomous cars to understand driving environments [2] and to detect overspeed in real time [3]. They are used to define the relationships between autonomous vehicles [4] and thus make it possible to regulate the circulation of intelligent vehicles. In [5], authors propose an ontology-based approach to facilitate the interpretation of the information collected by VANET (Vehicular Ad-Hoc Networks). The ontology developed is interpreted in each vehicle, thus facilitating communication between vehicles and communication between vehicles and infrastructure. In [6], the proposed approach, through an ontology, sets up a system making it possible to make driving decisions at uncontrolled intersections and on narrow roads. The approach differs from that presented in [2] by the possibility of down a complicated case into several situations and integrating the individual results to make a final decision. The approach

[2] https://repository.uneca.org/handle/10855/23406.

in [7] uses an ontology to associate the perceived information and its context in the interpretation of traffic scenarios and the estimation of the perceived entities in relation to the subject vehicle. The ontology developed is made up of three elements: mobile entities (vehicles, pedestrians), static entities (roads, intersections) and context parameters that connect two entities according to their state and the distance between them. One of the limits of this approach is that the ontology developed does not allow inference in any context but in the one for which it was defined. A model based on ontologies is also proposed in [4]. It allows autonomous vehicles to make decisions when faced with unusual situations. The ontology proposed in [8] makes it possible to encode the information collected by the sensors of a vehicle for their interoperability with the other actors of the ITS. This ontology involves four components: the accident, the environment, the vehicle and the occupant.

However, these solutions are not suitable in African countries where the context is different. In fact, most vehicles in these countries and infrastructure do not have intelligent technologies making thus communication between vehicles and their environment difficult. Added to this, the absence of concepts that can describe certain types of transport vehicles in ontologies presented above. Our proposal is at the level of data collection and exploitation by a platform in order to have a knowledge base on traffic accidents.

2.2 Ontological Needs

The aim of our research in this area is to provide a social and semantic web platform for describing and managing traffic accidents in Senegal. This work could be adapted in the context of other African countries having the same difficulties in traffic accidents management and road traffic regulation. The platform is based on an ontology of road accidents. With this ontology and the descriptions, inferences can be made in order to discover new knowledge related to accidents such as the main causes of accidents, the most accidental roads, drivers who commit more fatal accidents, etc.

Thus, in the platform for which the ontology is developed, traffic accidents can be described by three methods:

- manual description by firefighters, police, gendarmerie, witnesses;
- automatic description from images or videos with deep learning methods;
- semantic scraping in the online Senegalese press and social networks.

For illustration, we have collected an online press article of a road accident on Dakaractu[3], described by the following text:

> *"The public transport bus (Ndiaga-Ndiaye) commonly called "horaire" of Ndoffane which went to the weekly market of Birkelane (Louma) overturned in the early hours of this Sunday, March 8, 2020 near the village of Thiawando (Latmingué). The provisional assessment reports a death and several injured. Informed, the firefighters who deported to the scene of the accident, proceeded to their evacuation to the Kaolack Regional Hospital.".*

[3] https://www.dakaractu.com/Kaolack-Le-Horaire-de-Ndoffane-se-renverse-et-fait-1-mort_a184995.html.

This press article informs that the description of road accidents in Senegal is done in a context where existing ontologies cannot take into account in terms of vocabulary. Indeed, the use of local languages terms such as "Louma" in the description of accidents in the online Senegalese press and of urban language in social networks like "horaire" makes it difficult to use this information. Added to this is the ignorance of the traffic rules which are often different from those of developed countries. Faced with this need, the conception of an ontology of traffic accidents in Senegal finds its relevance. The following section presents the construction of this ontology.

3 Ontology Building

To construct the ontology, we adopt the methodology proposed in [12] even if other methods exist in the literature [9, 10] and [11]. This is an iterative ontology construction methodology that simplifies defining class hierarchies, properties of classes, instances and is adapted for declarative frame-based systems. It includes seven steps:

- **Step 1: Determine the domain and scope of the ontology**
  It is about determining the area that ontology is supposed to cover and reasons for which it will be used.
- **Step 2: Consider reusing existing ontologies**
  In this step, it is necessary to see if the system for which ontology is developed interacts with other applications requiring the reuse of existing ontologies or controlled vocabularies. In our case, we have identified the Road Accident Ontology[4] that will be used to identify animals that can be victims or causes.
- **Step 3: Enumerate important terms in the ontology**
  This step makes it possible to explicitly identify terms used and their meanings for a better understanding.
- **Step 4: Define the classes and the class hierarchy**
  Here, three methods are most used for defining classes in hierarchy. The first is the top-down process which stars with the definition of the most general concepts and then proceeds to their specialization. We did it with the concept Accident we specialized in Fatal_Accident, Non_Fatal_Accident and Material_Damage_Accident. The bottom-up process that starts with defining the most specific concepts and then grouping them together to find more general concepts is the second method. For example, the concepts Injured and Killed can have a superclass Victim. A combination of these two approaches is the third method.
- **Step 5: Define the properties of classes**
  Once classes defined, the internal structure of these classes must be found. This allows them to be linked with properties generally chosen from terms defined in step 3.
- **Step 6: Define the facets of the properties**
  This is to describe value type, allowed values, cardinality and other characteristics data properties can take.
- **Step 7: Create instances**
  The last step is to create instances of classes in the hierarchy.

[4] https://www.w3.org/2012/06/rao.html.

3.1 Domain Delimitation of Traffic Accident

An accident, as shown in Fig. 1, takes place on an infrastructure and involves one or more vehicles. The accident occurs according to circumstances and produces consequences such as material damage, injured who may be seriously or slightly injured or death. It takes place during a given period and can be relayed by a press article, a tweet, a video, a story told by a witness or a report established by a traffic officer. Infrastructure on which the accident occurs can be a road and is characterized in this case by traffic conditions (visibility, humidity, temperature, pressure, precipitation, wind, etc.). The vehicle involved in the accident is conducted by a driver with a license who commits an infringement and has passengers on board.

This description shows that several domains are involved in traffic accidents management. It's about:

- Infrastructure (roads, highways, tracks, intersections, roundabouts, tunnels, tolls, lights, signs, etc.). The ontology will be used to map all of the available transport infrastructure in the country.
- Security with the police and the gendarmerie responsible for ascertaining accidents and writing reports.

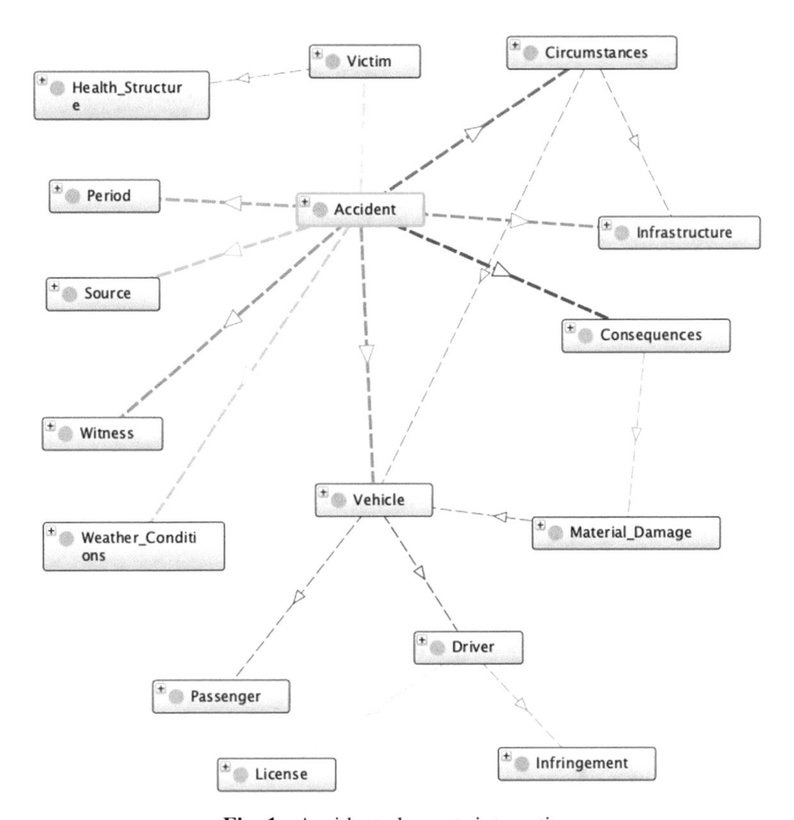

Fig. 1. Accident elements interaction

- Automobile with the modeling of all vehicles type.
- Health as the victims of accidents will have to be sent to health structures for their medical care.
- Insurance responsible for reimbursing damages
- Environment with climatic factors such as which play an important role in the upsurge in accidents (rain, wind, temperature, etc.).
- Media, especially the online press and social networks which serve as media for collecting information on accidents.
- Economy with the impact of accidents on the economy, particularly loss of life and property damage.
- Psychology with the modeling of the behavior of road uses (drivers, passengers, pedestrians, residents of the road).

3.2 Reusing Existing Ontologies

In our approach, we plan to study the influence of human behavior in accidents. These are first and foremost the drivers, but also passengers, pedestrians and animals that can lead drivers to commit infringements. Thus, we align Traffic Accident Ontology to Neuro Behavior Ontology (NBO)[5], an ontology that describes human and animal behavior and behavioral phenotypes. Behavior is defined in [13] to be the response of an organism

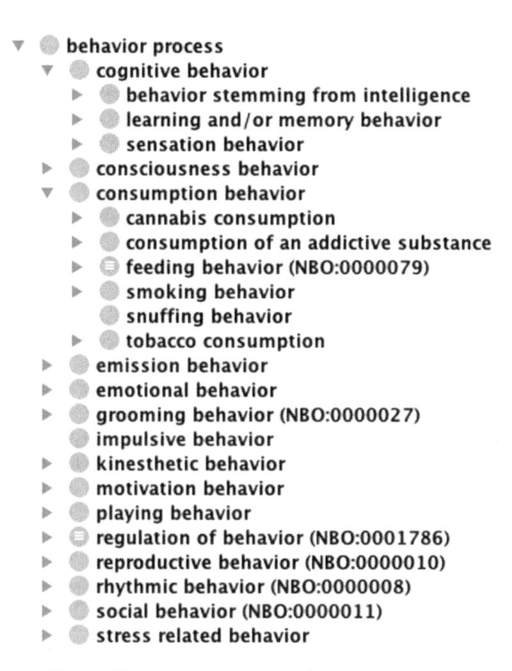

Fig. 2. Behavioral process hierarchy in NBO

[5] http://www.ontobee.org/ontology/nbo.

or a group of organisms to external or internal stimuli. The neurobehavior ontology [13] consists of two main components, an ontology of behavioral processes as shown in Fig. 2. And an ontology of behavioral phenotypes.

We will use in our system the Behavioral process to model driver movements in response to the lack of sleep or drug use. Indeed, the accident investigation reports on which we rely to record a traffic accident in Senegal mention the accident causes, in particular the faults committed by the driver. The investigation by security officers will reveal the driver's state of mind during an accident and the possible use of narcotic drugs.

In the ontology that we are going to design, the Driver concept will contain a behavioral attribute which describes the set of attitudes observed in a driver during the accident. This attribute is then used for matching at the NBO using an alignment algorithm.

3.3 Classes and the Class Hierarchy

The ontology must give more information about place as the objective in approach is the identification of areas at high risk of accidents. Figure 3. Indicates that an infrastructure can be an intersection, a road, a roundabout or a tunnel. A road is categorized as a tarred road (highways, national road, regional road, departmental road, municipal road, etc.) and untarred road.

Fig. 3. Infrastructures hierarchy

Traffic accident ontology specifies five family that are commercial vehicle, industrial vehicle, private vehicle, public transport vehicle and specific vehicle such as ambulance. Our conceptualization takes into account specific transport vehicles in Senegal such as the Ndiaga-Ndiaye, Car-Rapide and 7-place that are cargo transport vehicles transformed into passenger transport vehicles and the Clando which are non-regularized taxis. The specificity of our field of application means that terms from urban language are integrated into the ontology. Figure 4 Illustrates this classification.

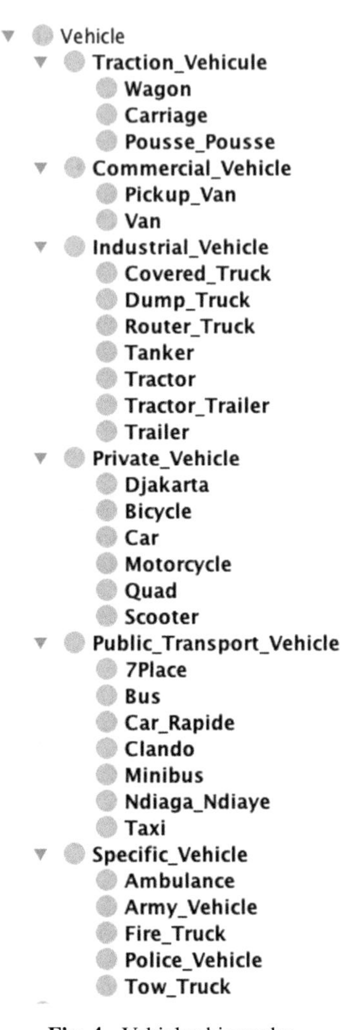

Fig. 4. Vehicles hierarchy

Traffic Accident Ontology also conceptualizes living beings whose action impacts in accidents. These are animals which can push the driver to make a bad maneuver, pedestrians and neighbors of the road who can be victims, passengers of the vehicle, witnesses of the accident, traffic officers in charge of recording accidents and driver as shown in Fig. 5.

In the next section, we formalize the ontology using a model. Model representation gives a clear specification of concepts, relationships, instances and axioms.

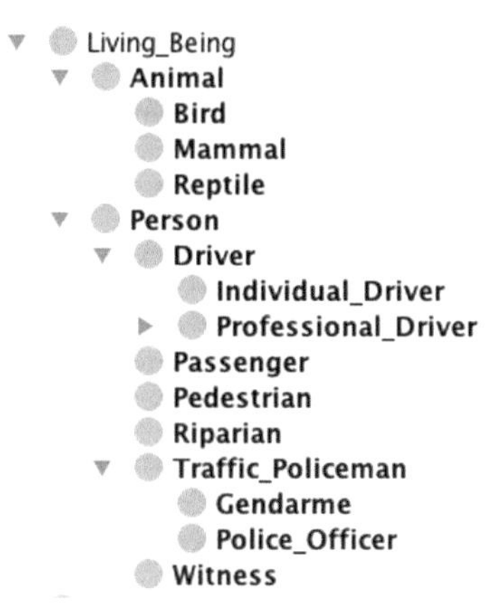

Fig. 5. Illustration of the living being classification involved in traffic accident

4 Ontology Formaling

According to Gruber, ontology defines a set of representational primitives with which to model a domain of knowledge or discourse. The representational primitives are typically classes, properties (attributes) and relations [14]. Such ontology, called light ontology, can be represented by a model O from [15] and [16] defined as following:

$$O = \{C, R, A, T, CAR_R, H^C, \sigma_R, \sigma_{CARR}, \sigma_A\}$$

where

- C represents all the concepts of ontology;
- A, all attributes;
- T, a set containing the types of attribute;
- $R \subseteq (C \times C)$ defines associative relations between concepts;
- $H^C \subseteq (C \times C)$ defines hierarchy of concepts;
- CAR_R represents the associative relation R characteristic which can be symmetrical, transitive, reflexive,...;
- σ_R: $R \rightarrow C \times C$ defines the signature of the associative R;
- σ_{CARR}: $R \rightarrow CAR_R$ is the signature of the relation specifying characteristic of the relation R. We note an associative relation R_k transitive by signature $\sigma_{CARR}(R_k$, Transitive).;
- σ_A: $A \rightarrow C \times T$ defines the signature of the attribute relation between a concept and an attribute.

Example: Consider the accident elements interaction in Fig. 1. The corresponding ontology formalization is defined as followed:

- C = {Accident, Vehicle, Victim, Witness, Damage, Weather_Conditions, Health_Structure, Period, Traffic_Policeman, Driver, License, Infringement, Passenger, Infrastructure, Traffic_Conditions}

- R = {causes, commits, engenders, is_evacuated, has, having, contains, involves, is_driven_by, is_noted_by, is_occurred_during, is_occurred_under, occurs_in_front_of, takes_place_on}
- A = {category, city, description, coordinates, place, visibility, start_time, end_time, age}
- T = {xsd:string, xsd:int. Xsd:decimal, xsd:dateTime}
- S_{HC} = {H^c(Fatal_Accident, Accident), H^c(Non_Fatal_Accident, Accident), H^c(Material_Damage_Accident, Accident), H^c(Injured, Victim), H^c(Killed, Victim), H^c(Hospital, Health_Structure), H^c(Health_Center, Health_Structure), H^c(Health_Post, Health_Structure)}
- $S_{\sigma R}$ = {σ_R(Accident, causes, Victim), σ_R(Driver, commits, Infringement), σ_R(Vehicle, contains, Passenger), σ_R(Victim, is_evacuated, Health_Structure), σ_R(Driver, has, License), σ_R(Infrastruture, having, Traffic_Conditions), σ_R(Accident, involves, Vehicle), σ_R(Vehicle, is_driven_by, Driver), σ_R(Accident, is_noted_by, Traffic_Policeman), σ_R(Accident, is_occurred_during, Period), σ_R(Accident, is_occurred_under, Weather_Conditions), σ_R(Accident, occurs_in_front_of, Witness), σ_R(Accident, takes_place_on, Infrastruture)}
- $S_{\sigma A}$ = {σ_A(License, category, xsd:string), σ_A(Infrastruture, city, xsd:string), σ_A(License, delivery_year, xsd:int), σ_A(License, duration, xsd:int), σ_A(Accident, description, xsd:string), σ_A(Accident, place, xsd:string), σ_A(Accident, latitude, xsd:decimal), σ_A(Accident, longitude, xsd:decimal), σ_A(Weather_Conditions, visibility, xsd:string), σ_A(Period, start_time, xsd:dateTime), σ_A(Period, end_time, xsd:dateTime), σ_A(Victim, age, xsd:int)}
- $S_{\sigma CARR}$ = {σ_{CARR}(causes, Asymmetric), σ_{CARR}(causes, Irreflexive), σ_{CARR}(commits, Asymmetric), σ_{CARR}(commits, Irreflexive), σ_{CARR}(engenders, Asymmetric), σ_{CARR}(engenders, Irreflexive)}

5 Conclusion

In this paper, we construct a domain ontology on traffic accident for identifying areas at high risk of accidents in less developed countries. We first built using an iterative construction method. We then proposed a model for formalizing the ontology. Our proposal made it possible to have structured and semantic data on traffic accidents and also helped to resolve difficulties linked to traffic accidents management.

In the future work, we continue to populate the ontology with data from the Senegalese online press and police investigation reports which will allow an inference in any context of traffic accident. In a second step, the populating of the ontology will be done automatically with images recognition taken during accident. We will investigate the Graphs Theory including Bayesian Networks, Markov Chains, Propagation of beliefs among others for reasoning on the sanctions to be applied to drivers at fault.

References

1. World Health Organization, Global status report on road safety 2018, Switzerland

2. Zhao, L., et al.: Ontologies for advanced driver assistance systems. In: The 35th Semantic Web & Ontology Workshop (SWO) (2015)
3. Zhao, L., et al.: An ontology-based intelligent speed adaptation system for autonomous cars. In: Supnithi, T., Yamaguchi, T., Pan, J., Wuwongse, V., Buranarach, M. (eds.) Semantic Technology. JIST 2014. LNCS, vol. 8943. Springer, Cham (2015). https://doi.org/10.1007/978-3-319-15615-6_30
4. Pollard, E., Morignot, P., Nashashibi, F.: An ontology-based model to determine the automation level of an automated vehicle for co-driving. In: Proceedings of the 16th International Conference on Information Fusion. IEEE (2013)
5. Gao, H., Liu, F.: Estimating freeway traffic measures from mobile phone location data. Eur. J. Oper. Res. **229**, 252–260 (2013)
6. Zhao, L., et al.: Ontology-based driving decision making: a feasibility study at uncontrolled intersections. IEICE Trans. Inf. Syst. **100**(7), 1425–1439 (2017)
7. Armand, A., Filliat, D., Ibañez-Guzman, J.: Ontology-based context awareness for driving assistance systems. In: 2014 IEEE Intelligent Vehicles Symposium Proceedings, 2014, pp. 227–233 (2014)
8. Barrachina, J., et al.: VEACON: a vehicular accident ontology designed to improve safety on the roads. J. Netw. Comput. Appl. **35**(6), 1891–1900 (2012)
9. Uschold, M., King, M.: Towards a methodology for building ontologies. In: Skuce, D. (Ed.), IJCAI'95 Workshop on Basic Ontological Issues in Knowledge Sharing, pp. 6.1–6.10, Montreal, Montreal, Canada (1995)
10. Uschold, M., Grüninger, M.: Ontologies: principles, methods and applications. Knowl. Eng. Rev. **11**(2), 93–155 (1996)
11. Staab, S., Schnurr, H.P., Studer, R., Sure, Y.: Knowledge processes and ontologies. IEEE Intell. Syst. **16**(1), 26–34 (2001)
12. Noy, N.F., McGuinness, D.L.: Ontology development 101: a guide to creating your first ontology (2001)
13. Gkoutos, G.V., Schofield, P.N., Hoehndorf, R.: The neurobehavior ontology: an ontology for annotation and integration of behavior and behavioral phenotypes. In: International Review of Neurobiology, vol. 103, pp. 69–87. Academic Press (2012)
14. Gruber, T.: Ontology. In: Binder, M.D., Hirokawa, N., Windhorst, U. (eds.) Encyclopedia of Neuroscience. Springer, Berlin, Heidelberg (2009). https://doi.org/10.1007/978-3-540-29678-2_4225
15. Sall, O., et al.: A model for ripple effects of ontology evolution based on assertions and ontology reverse engineering. In: García-Barriocanal, E., Cebeci, Z., Okur, M.C., Öztürk, A. (eds.) Metadata and Semantic Research. MTSR 2011. CCIS, vol. 240. Springer, Berlin, Heidelberg (2011). https://doi.org/10.1007/978-3-642-24731-6_15
16. Gaye, M., et al.: Measuring inconsistencies propagation from change operation based on ontology partitioning. In: 11th IEEE International Conference on Signal-Image Technology & Internet-Based Systems (SITIS) (2015)

Ontology Partitioning for Managing Change Effects

Mouhamadou Gaye(✉) and Ibrahima Diop

Computer Science and Engineering for Innovation Laboratory, Assane Seck University, Ziguinchor, Senegal
{m.gaye,ibrahima.diop}@univ-zig.sn

Abstract. Ontology evolution refers to the process of an ontology modification in response to change in the domain or its conceptualization. It takes in several phases which the most important are changes representation, semantics of changes, propagation and validation. Analysis of changes to be made in ontology is necessary to identify potential consequences on ontology and on dependent objects. Indeed, a modification of an ontological entity can generate impacts making the system inconsistent. Thus, impacts propagation is important to keep the system stable. However, managing changes and their effects is not a simple task; it is more difficult if concerned ontology is voluminous.

We propose an approach to partition large ontology for change effects managing. Our proposed approach consists of creating a weighted dependency graph from ontology structure and then determining communities using the Island Line algorithm.

Keywords: Ontology · Evolution · Change Impacts · Partitioning

1 Introduction

The SemanticWeb, assigned to Tim Berners-Lee [1] of the World Wide Web Consortium (W3C), is an infrastructure whose purpose is to enable software agents to more effectively help different types of users in their access to resources on the web [2]. This mission is accomplished through the use of ontologies whose primary purpose is to provide a shared vocabulary for a given area of knowledge. However, these ontologies are constantly evolving to readjust to the needs for which they were defined. Evolution refers to the process of modification in response to a change in domain or its conceptualization. This evolution must, according to Stojanovic and al. in [3], make it possible to solve the changes made in ontology, and to guarantee the consistency of concerned ontology and of all dependent objects. Thus, it is then necessary to manage impacts of an ontological entity change before beginning any process of evolution. Maintaining large ontologies such as NCI-Thesarus and Gene Ontology is all the more complex [4] because it is often entrusted to groups of experts who only support the part of ontology they have created. Another problem related to ontology evolution is the tracking inconsistencies propagation on the entities of the modified ontology. Existing work does not adequately address this aspect no less important.

T. Matsuo et al. (Eds.): AIMD 2019, LNNS 677, pp. 132–146, 2023.
https://doi.org/10.1007/978-3-031-30769-0_13

In this article, we propose an approach to partition large ontology for change effects managing. Our proposed approach consists of creating a weighted dependency graph from ontology structure and then determining communities using the Island Line algorithm.

The rest of the article is organized as follows. We present in Sect. 2 a survey on ontology evolution and change propagation. In Sect. 3, we give basic elements on which our proposals relate. Section 4 is devoted to ontology partitioning method that we proposed and we end with a summary of our work and its prospects.

2 Literature Survey

Ontology evolution refers to the process of modifying ontology in response to a certain change in domain or in its conceptualization [5]. For Stojanovic and al. in [3], ontology evolution is the appropriate modification of ontology and the consistent propagation of changes in the dependent artefacts, that is to say referenced objects, ontologies dependent and software applications using the ontology. According to this article, evolution must (1) make it possible to solve the changes occurred in the ontology, (2) guarantee the consistency of concerned ontology and of all the dependent objects, (3) be supervised allowing users to process changes more easily; and (4) provide users with guidance for continual refinement of the ontology. Study in [6] makes distinction between ontology evolution, ontology management, ontology modification and ontology versioning by giving definitions below.

- Ontology management is the set of methods and techniques necessary for an efficient use of multiple variants of ontology coming from different sources and intended for different tasks. Thus, ontology management system is a platform for creating, modifying, managing versions, querying and storing ontologies.
- Ontology modification is adapted when ontology management system makes it possible to execute changes in ontology without considering consistency.
- Ontology versioning adapts to situation where ontology management system allows the processing of changes by creating and managing different versions of ontology.
- Ontology evolution as for it, is appropriate when ontology management system facilitates modification of an ontology while preserving its consistency.

In any case, changes made to ontology can be explained by environment evolution for which it has been developed. Many work on ontology evolution are based on version logs. Klein and Fensel in [7], discuss problem of ontology versioning especially on the identification and specification aspects of changes. They point out that there are three methods for linking ontology versions with data sources: using (1) ontology with the data source, (2) a version of the more recent ontology, and (3) an older version. Method in [8] makes it possible to compare versions of ontology and specify their conceptual relations. To visualize differences, this approach use a rule-based mechanism and classify RDF ontology changes. The study emphasizes that it is not possible to automatically determine whether a change is a conceptual change or an explanatory change.

In this category of approaches based on version logs, work from [9] gives a formalization of elementary and complex changes made to a version V_n of an ontology to obtain another version V_{n+1}. The formalism is semi-compatible with OWL and makes history of changes interoperable and shareable. Method in [10] describes possibilities for automatic identification of changes between ontologies using heuristics. Heuristics are also used in PromptDiff algorithm [11] to compare ontology versions. Approach in [12] describe through techniques and heuristics, different ways of representing information of changes. However, definition of heuristics to have complex changes from simple changes is not rigorously established. In [13], proposed method allows to detect and represent relations between versions of ontology in three stages: preliminary detection of mapping, refinement and cleaning of the mapping and computation of the inverse mapping. Evolution of logs is also used in [14] to draw lines of axioms and annotations by associating each axiom or annotation with a set of metadata. Idea is to study individual evolution of axioms or annotations instead of evolution of the entire ontology. Approaches based on logs version, widely treated in the literature, nevertheless pose limits residing on language used, non-support of dependencies, absence of a collaborative development environment and non-coverage of all the aspects of evolution [15].

Some studies deal ontology evolution with databases evolution methodologies [16, 17]. Noy and Klein in [18] study difference between ontology evolution and database schemas evolution. For them, traditional distinction between evolution and versioning is not applicable to ontologies. Indeed, change operations for ontologies are different from those of databases and semantics are more explicit with ontologies. Nevertheless, a method of storing multiple versions of ontology based on relational databases, proposed in [16], allows through SQL queries to make structural comparisons of ontologies. However, the model does not support Boolean expressions, and ontology definitions are decomposed and converted to relational tables according to ontology metadata. Approaches based on change patterns exist in [19] and [20]. Patterns correspond to change, inconsistency and alternative resolution dimensions. Based on these patterns and links between them, the approach defines an automated process to drive application of the changes. There are also ontology evolution approaches using descriptive logic in their model. This is the case of approach in [21] which defines a temporal logic metric with modularities and a hybrid satisfaction operator for analysing evolution. Though, domain restriction or extension or co-domain operations are not taken into account. Other models use graph grammars to formalize and man age evolution of ontologies. In [22], an ontological change management approach based on the grammar of graphs is proposed. A grammar of graph [23] is a pair composed of an initial graph and a set of production rules. The approach defines a rule by a left subgraph representing precondition and a right subgraph representing post-condition of the rule. Work in [24] based on revision of beliefs [25], presents a general framework for the representation of uncertainties, including the theory of probability as a special case.

Nonetheless, ontology evolution is intrinsically linked to measurement and propagation of inconsistencies induced by a modification operation in ontology. Thus, many ontology evolution approaches are concerned with inconsistencies measurement or propagation.

In [26], Ticky presents a change propagation scenario in software that extracts information from software architecture to build a basic version of the program. The Intelligent recompilation method proposed uses dependencies between system components to suggest elements that depend on a previously modified component. Intelligent recompilation provides, according to [27], a highly specialized change propagation algorithm for a bottom-up propagation of an entity E_j to an entity E_i. Propagation is modeled in [28] as a sequence of snapshots and each snapshot represents a particular moment in the process. Two processes are presented in this proposal: change-and-fix and top-down propagation. Changeand- fix process first defines the types of dependencies between program components in inbound dependencies and outbound dependencies. The algorithm then consists in marking entities in dependence on modified entity, that is to say entities located in the neighborhood of the modified entity. Top-down propagation process is an improved version of the first process that ensures absence of cyclic dependencies. Proposal in [29] proceeds by derivation of changes to analyze and follow impacts of ontological modifications.

All these approaches deal with inconsistencies measurement and propagation in a knowledge base. However, they do not take into account the size of the knowledge base to which the estimations relate. For a large ontology of the size of Gene Ontology, these propositions do not make it possible to effectively follow propagation of these inconsistencies in the other entities of the ontology. Our work in this article fits in this direction.

3 Preliminaries

3.1 Ontological Changes

According to the taxonomy criterion considered, ontological changes are differently classified in the literature. Klein and al. in [30] enumerate two types of changes [31] that affect ontological conceptualization: conceptual changes and explicative changes that are changes in the specification of the conceptualization and not in conceptualization. Klein adds a third type of change in [32] and is now talking about change in conceptualization, change in specification, and change in representation. Regarding detection, ontological changes can be classified in two categories: top-down changes and bottom-down changes [3]. Top-down changes are explicit changes, introduced, for example, by a top-manager who wants to adapt system to new requirements and can be easily realized by the ontology evolution system. Bottom-up changes are extracted from all instances of ontology. For Stojanovic [6], ontological changes form two families: elementary changes also called simple or atomic changes and complex changes that do not represent predominance in the proposed research results.

An elemantary change is a change in ontology that modifies (adds or delete) a single entity of the ontological model [6]. This entity can be a concept, a relation, an instance, a type or a relation characteristic. A list of simple ontological changes is mentionned in [3, 6, 32, 33] and [34] A complex change is an ordered collection of elemental changes [9]. For many authors, there is no difference between complex change and composite change. Stojanovic nevertheless makes this difference and defines a composite change as an ontological change that modifies (creates, renames, changes) the neighborhood of an ontological entity and a complex change as a change that can be decomposed into a combination of at least two elementary or composite changes [6]. Complex changes are cited in [3, 6, 32] and [9].

3.2 Ontology Model Formalism

Ontology defines a set of representational primitives that models a knowledge domain. Representational primitives are typically classes, attributes, and relations. The representation of these primitives refers to the use of a representation language that clearly specifies concepts, relations, instances and axioms.

3.2.1 Lightweight Ontology Formal Model

A lightweight ontology is an ontology that does not include in its definition the use of axioms, as opposed to a heavy ontology. It is a collection of concepts and attributes linked by subsumption, object properties, and datatype properties. We therefore represent a lightweight ontology by a structure O, derived from model established in [35] and defined as follows:

$$\mathrm{O} = \{\mathrm{C}, \mathrm{R}, \mathrm{A}, \mathrm{T}, \mathrm{CAR}_R, \mathrm{H}^C, \mathrm{H}^R, \sigma_R, \sigma_{CARR}, \sigma_A, \sigma_T\}$$

with:

- C, A, T, CARR are respectively sets containing, the concepts of ontology, the relations of attribute, the types of attribute and characteristics of associative relations;
- $\mathrm{R} \subseteq (\mathrm{C} \times \mathrm{C})$ is associative relations set. It makes it possible to define the semantic types of relations connecting the concepts of ontology in $(\mathrm{C} \times \mathrm{C})$;
- $\mathrm{H^C}$ hierarchy (taxonomy) of concepts: $\mathrm{H}^C \subseteq (\mathrm{C} \times \mathrm{C})$, $\mathrm{H}^C(\mathrm{C}_i, \mathrm{C}_j)$ means that C_i is a sub-concept of C_j, for subsumption relations between ontology concepts;
- $\mathrm{H^R}$ hierarchy of relations: $\mathrm{H}^R \subseteq (\mathrm{R} \times \mathrm{R})$, $\mathrm{H}^R(\mathrm{R}_i, \mathrm{R}_j)$ means that R_i is a sub-property of R_j, for subsumption between ontology properties;
- σ_R: $\mathrm{R} \rightarrow \mathrm{C} \times \mathrm{C}$ is the signature of an associative relation. We will note $\sigma_R(\mathrm{C}_i, \mathrm{R}_k, \mathrm{C}_j)$ the signature of the associative relation R_k between the concepts C_i and C_j;
- σ_A: $\mathrm{A} \rightarrow \mathrm{C} \times \mathrm{T}$ is the relation of attribute signature, T is composed of the simple types. It is noted as $\sigma_A(\mathrm{C}_i, \mathrm{A}_k, \mathrm{T}_j)$ specifying the relation of attribute between a concept C_i and a A_k attribute having values of the type T_j;
- σ_T: $\mathrm{A} \rightarrow \mathrm{T}$ is the signature of the relation associating with an attribute A_k, the T_j type in the form σ_T $(\mathrm{A}_k, \mathrm{T}_j)$ specifying that the $\mathrm{A_k}$ attribute is associated with values of the T_j type;

- σ_{CARR}: R $\rightarrow$ CAR_R is the relation specifying the characteristic of an associative relation. We will, thus, note an associative relation R_k transitive by signature $\sigma_{CARR}(R_k,$ Trans).

Ontology formal model is not always easy to exploit, especially when ontology consists of a very large number of entities. Thus, we resort to graphical representation, more appropriate to certain situations all the more since it allows an ontology simple interpretation.

3.2.2 Graph Ontology

Definition 1: Graph ontology

An ontological graph is a tuple G = (E, Γ) where E is a set of entities which can be concepts or types and Γ an application from E to P(E) where P(E) contains all the set included in E and:

$$\Gamma \in \{H^C, \sigma_R, \sigma_A\}$$

A tuple (E_i, E_j) such that $E_j \in \Gamma(E_i)$ denotes a labeled edge between E_i and E_j where E_i is the source node and E_j the target node. With this graphical representation, the determination of dependencies between entities on ontology becomes easier. However, for a voluminous ontology, manipulation is not simple enough. To overcome this difficult, we use partitioning to manage evolution.

3.2.3 Ontology Dependencies

A dependency represents a relation, a link, an association between two objects or two groups of objects. Ontology components constitute entities linked by several dependency relationships.

Definition 2: Concepts dependency

Two concepts C_i and C_j are directly depend if C_i and C_j are linked by a subsomption or a associative relation. In the other words:

$$\text{dependency}(C_i,\ C_j)\ \text{if} \exists\ H^C \Big| H^C(C_i,\ C_j)\ \text{or if} \exists \sigma_R(C_i, R_k, C_j)$$

where R_k is an associative relation.

Definition 3: Concept and type dependency

A concept C_i and a type T_j are directly depend if C_i and T_j are linked by an attribute relation, that is:

$$\text{dependency}(C_i, T_j)\ \text{if} \exists\ \sigma_A | \sigma_A(C_i, A_k, T_k)$$

with A_k an attribute relation.

Dependencies between ontological entities are necessary to determine impact flows used in inconsistencies measurements definition and ontology partitioning method.

4 Ontology Partitioning

4.1 Ontology Partitioning Object

Large ontologies are noted in medicine and agronomy with ontologies such as AGROVOC[1], NALT[2], NCI[3], YAGO[4], Gene Ontology[5]. YAGO (Yet Another Great Ontology) ontology is a huge knowledge base, developed at the Max Planck Institute of Computer Science in Saarbrücken, describing more than 2 million entities of people, organizations, cities and towns, containing more than 20 million facts about these entities. Its data comes from Wikipedia[6] and is structured using WordNet[7]. Gene Ontology purpose to provide a controlled vocabulary that can be applied to all organisms, such as a knowledge base of genes and protein roles in cells [36]. The initial goal was to allow researchers to query related databases and obtain related proteins and gene products. Ontology has tens of thousands of concepts.

However, maintaining these large ontologies is a difficult task. If an update takes place on an entity of the ontology, the follow-up of its impacts on the whole ontology requires quite complex techniques. One of the techniques uses ontology decomposition into sub ontologies using partitioning methodologies. Ontology partitioning can be considered as problem of grouping entities of ontology (concepts, instances, and types) into clusters or communities so that the entities in a community share more properties within the community than they do outside. It was mostly used for ontology alignment, but it is also used in social networks, biological networks, information networks, language networks, and so on.

Grahne and Räihä, in their work on database schema decomposition [37], attest that good partitioning must respect the principles of:

- minimal redundancy: partitioning must represent significant objects;
- representation: it must be possible to recover the universal relation of the component relations;
- separation: when a relation is updated, it must be possible to ensure that dependencies in the universal relation always hold after the update, without actually building the universal relation.

In context of ontology partitioning, these principles are also relevant. A good ontology decomposition must represent all significant entities of the ontology, cover the ontology (the union of all the partitions must give the initial ontology) and preserve the initial dependencies.

---

[1] https://agrovoc.fao.org/.
[2] http://agclass.nal.usda.gov/agt.shtml.
[3] https://bioportal.bioontology.org/ontologies/ncit.
[4] http://www.mpi-inf.mpg.de/departments/databases-and-informationsystems/research/yago-naga/yago/downloads.
[5] http://geneontology.org/.
[6] https://en.wikipedia.org.
[7] https://wordnet.princeton.edu.

Most of decomposition approaches in the literature are focused for ontologies alignment. Partitioning methodology in [38] is guided by the existence of two ontologies to be aligned. The tool TaxoPart thus developed uses a hierarchical classification algorithm [39] which iteratively gathers in the same block, close concepts according to a similarity measure relying only on the relative position of the concepts in ontology. The tool breaks down the first most structured ontology and forces the second to follow this decomposition. In our context, this methodology is not appropriate because we have only one ontology that we seek to divide. Work in [40] presents two decomposition approaches based on graph partitioning algorithms. First approach consists in determining a minimal separator line of the graph in two disjoint subsets. Second approach is inspired by the image segmentation methodology. Approach in [41] proposes a partitioning method based on the concepts hierarchy. It consists of creating a weighted dependency graph from ontology structure and then determining modules using the Island Line algorithm [42]. Our partitioning method follows this same process but differs by the definition of the weights and dependencies types between ontological entities.

4.2 Ontology Partitioning Algorithm

Partitioning methodology we propose in our work is inspired by approach proposed in [41]. It consists of creating a weighted dependency graph from ontology structure and then determining communities using the Island Line algorithm. Ontology is a set of entities consisting of concepts, instances, and types related by subsumption relations, associative relations, and attribute relations. For each type of relation between two entities E_i and E_j, we assign a value P_{ij} according to direction of the propagation flow impacts.

- If $H^C(E_i, E_j)$ then $P_{ji} = 1$ and $P_{ij} = 0$;
- If $\sigma_R(E_i, R_k, E_j)$ then $P_{ji} = 1$ and $P_{ij} = 0$;
- If $\sigma_A(E_i, A_k, T_j)$ then $P_{ji} = 1$ and $P_{ij} = 0$.

This value is used to define dependency weight between two entities of the ontological graph.

Definition 4: Dependency weight

Let $G = (E, \Gamma)$ be an ontological graph, E_i and E_j two entities of E. We define the weight of the dependency between E_i and E_j as follows:

$$w(E_i, E_j) = \frac{P_{ij}+P_{ji}}{\sum_{k=1}^{N} (P_{ik}+P_{ki})} \tag{1}$$

N is the number of entities to which E_i is connected in G.

This weight will be used in the algorithm for communities' detection on ontology.

Definition 5: Ontological dependency graph

An ontological dependency graph is a triplet $G_d = (E, \Gamma, W)$ where E is a set of entities that can be concepts or types, Γ an application of E to the set P(E) of the parts of E and W a function of dependency between two entities of E. Figure 1 shows example of an ontological dependency graph.

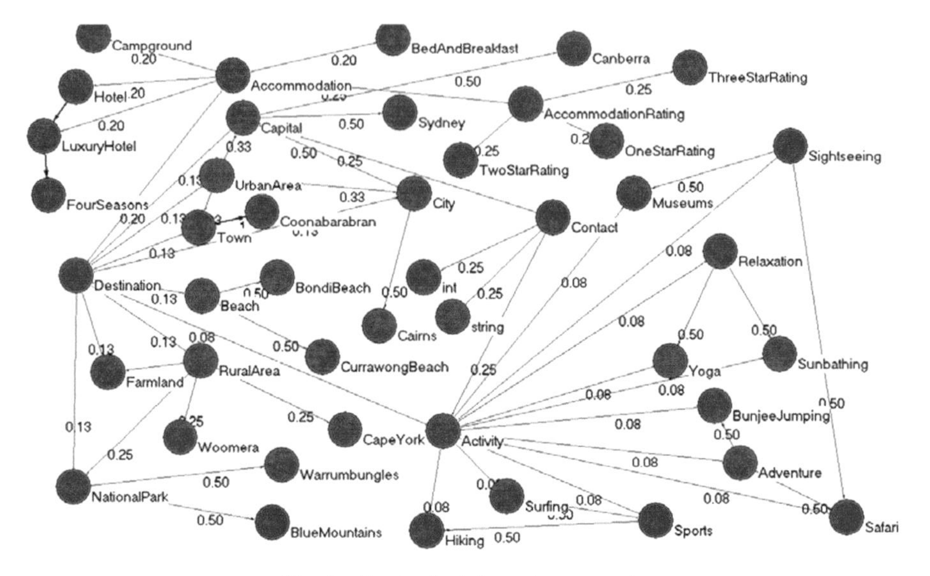

Fig. 1. Ontological dependency graph

Definition 6: Ontological community

Let $G_d = (E, \Gamma, W)$ an ontological dependency graph. A community $C \subseteq G_d$ is a set of concepts, instances or types such that two entities of C share more properties inside C than outside.

Ontology partitioning approach we propose uses the Island Line algorithm to decompose ontology. The algorithm determines the minimum separator line as a decomposition criterion and the number of separation lines is variable and depends on ontology size.

Definition 7: Edge island

A set of nodes V is an edge island if:

- It is a singleton or;
- The subgraph corresponding is a connected graph such that:

$$\max_{E_i \in V \wedge E_j \notin V} w(E_i, E_j) \leq \min_{E_k, E_l \in V} w(E_k, E_l) \#(2) \tag{2}$$

Edge island $V \subseteq G$ is regular edge island, if stronger condition holds:

$$\max_{E_i \in V \wedge E_j \notin V} w(E_i, E_j) < \min_{E_k, E_l \in V} w(E_k, E_l) \#(3) \tag{3}$$

4.3 Principle of the Algorithm

Ontology partitioning algorithm begins with a division of the ontological graph into subsets called islands. Each island is determined by its port designating the nearest point of the island. This algorithm creates first a set of islands grouped under a new sub-island name.

Second step of the algorithm is to choose among created islands those whose number of elements (number of vertices) is between the minimal and maximal (initially fixed) as candidates for partitioning. The islands whose number of entities is lower than the minimum fixed are removed, those whose number is greater than the maximum fixed are subdivided into islands that will be integrated to all islands.

The third and final step determines the partitions by taking as inputs all the candidates. Each candidate is first expanded with the nearest isolated vertices based on their weight before being decomposed.

We consider that an entity belongs to one and only one community. If an entity must follow two communities by the affinity of the shared properties, it is related to the one of which it is closest.

4.4 Algorithm Complexity

The partitioning algorithm we proposed makes an in-depth run of an ontological graph and takes all possible vertex pairs to determine the candidates. So with a graph of n vertices, we will have C_n^2 combinations, which gives:

$$C_n^2 = \frac{n!}{2!(n-2)!} = \frac{n(n-1)}{4} = \frac{n^2-n}{4} \quad (4)$$

Thus, the complexity of the algorithm is expressed in $\Theta(n^2)$. It is quadrastic while it is shown in [43] that the reasoning algorithms on modular ontologies are of exponential complexity (Fig. 2).

Algorithm 1: Ontology partitioning
Inputs: G = (E, Γ, W), a dependency ontological graph and maxCties, maximum communities number to obtain.
Outputs: numberCties, communities number obtained.

```
1  min = 1
2  max = |E| - 1
3  islands = {{v} : v ∈ E}
4  subIslands = ∅;
5  Order E in dercroissant according to the dependency weight w
6  For (u, v) ∈ G do
7        i1 = island ∈ islands : u ∈ island
8        i2 = island ∈ islands : v ∈ island
9        If (i16≠ i2) then
10             island = createIsland()
11             island.port = u
12             island.subIsland1 = i1
13             island.subIsland2 = i2
14             subIslands = subIsland ∪ {island} \ {i1, i2}
15       EndIf
16 EndFor
17 candidates = ∅
18 While (sousIles≠∅) do
19       Select island ∈ subIslands
20       subIslands = subIslands \ {island}
21       If (|island| ≤ min) then
22             Delete island
23       Else if (|island| ≥ max) then
24                   subIslands = subIslands ∪ {island.subIsland1, island.subIsland2}
25                   Delete island
26             Else
27                   candidates = candidates ∪ {island}
28             EndIf
29       EndIf
30 EndWhile
31 For community ∈ candidats do
32       expand(community, |E|)
33       partitionnner(maxCties, community, numberCties)
34 EndFor
```

Algorithm 2: Extend community
Inputs: a community C, set of entities E and set of candidates.
Outputs: community C eventually enriched

```
1  For v ∈ E : v is separated do
2        If (w(v, C.port) ≤ min_{Ci∈candidates} w(v, Ci.port) et |C| ≤ |E|) then
3              C = C ∪ {v}
4        EndIf
5  EndIf
```

Algorithm 3: Partition community
Inputs: a community C, maxCties, maximum communities and the set of candidates.
Outputs: numberCties, number of communities

1 **If**(numberCties ≤ maxCties) **then**
2 $C_1 = \{u\}: u \in C$
3 $C_2 = \{v\}: v \in C$
4 **For** $E_i \in C$ **do**
5 **If** ($w(E_i, C_1.port) < w(E_i, C_2.port)$) **then**
6 $C_1 = C_1 \cup \{E_i\}$
7 **Else**
8 $C_2 = C_2 \cup \{E_i\}$
9 **FinSi**
10 **FinPour**
11 Delete C
12 numberCties = numberCties + 1
13 candidates = candidates $\cup$ $\{C_1, C_2\}$
14 **EndIf**

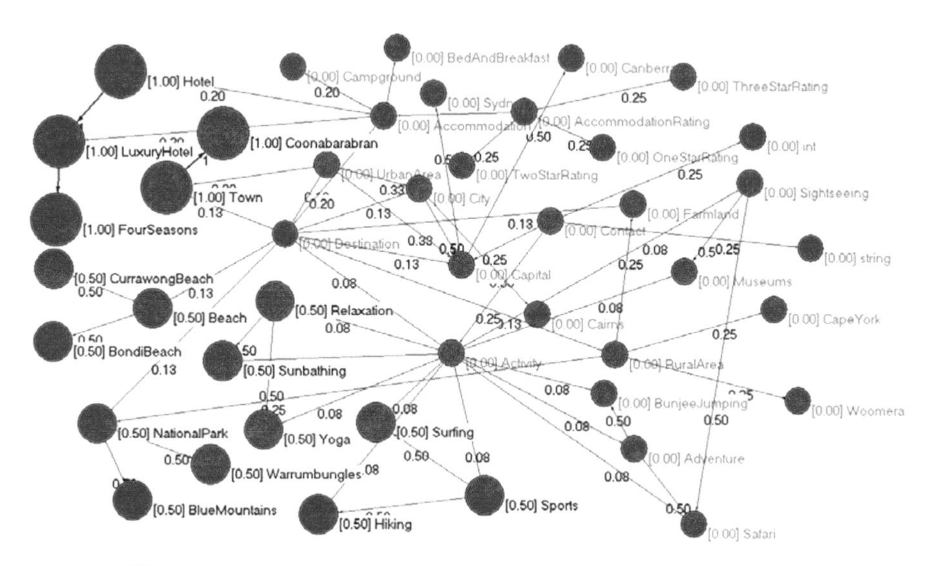

Fig. 2. A partition into three communities of the ontology given in Fig. 1

5 Conclusion

Work in this paper has consisted of partitioning large ontology to manage change effects in ontology evolution. The first step was to define ontological changes and ontology model formalism, then, in a second step, to develop our approach which decomposes large ontology into communities from the weighted dependency graph of the ontology.

This work has made it possible to set up a management system for the evolution of large ontologies widely used in the field of health, biology and especially agriculture.

However, the results can still be improved to take into account the management of ontology evolution in the context of data integration by a mediator system. This opens up several research perspectives that we develop in our next contributions. We have discovered, as part of the research, relevant research issues related to our work. The measurement and propagation of inconsistencies in the context of an ontology-based integration system is an interesting prospect to explore.

Another avenue of research to explore is the inconsistency resolution. The existing approaches bring the beginnings of solution but pain, especially in the case of a voluminous ontology or an integration system based on ontologies. We believe that the Markov decision process could be promising.

References

1. Berners-Lee, T., Hendler, J., Lassila, O.: The semantic web. Sci. Am. **285**(5), 28–37 (2001)
2. Laublet, P., Reynaud, C., Charlet, J.: Sur quelques aspects du Web sémantique. Assises du GDRI3, pp. 59–78 (2002)
3. Stojanovic, L., Maedche, A., Motik, B., Stojanovic, N.: User-driven ontology evolution management. In: Gómez-Pérez, A., Benjamins, V.R. (eds.) Knowledge Engineering and Knowledge Management: Ontologies and the Semantic Web. EKAW 2002. LNCS, vol. 2473, pp. 285–300. Springer, Berlin, Heidelberg (2002). https://doi.org/10.1007/3-540-45810-7_27
4. Ahmed, S.S., Malki, M., Benslimane, S.M.: Ontology partitioning: clustering based approach. Int. J. Inf. Technol. Comput. Sci. (IJITCS) (2015)
5. Flouris, G., Manakanatas, D., Kondylakis, H., Plexousakis, D., Antoniou, G.: A classification of ontology change. In: SWAP (2006)
6. Stojanovic, L.: Methods and Tools for Ontology Evolution. Université de Karlsruhe (2004)
7. Klein, M.C., Fensel, D.: Ontology versioning on the semantic web. In: SWWS, pp. 75–91 (2001)
8. Klein, M., Fensel, D., Kiryakov, A., Ognyanov, D.: Ontoview: comparing and versioning ontologies. Collected Posters ISWC (2002)
9. Codruta Rogozan, D.: Gestion de l'évolution des ontologies: méthodes et outils pour un référencement sémantique évolutif fondé sur une analyse des changements entre versions d'ontologie. Université du Québec à Montréal (2008)
10. Tury, M., Bieliková, M.: An approach to detection ontology changes. In: Workshop Proceedings of the Sixth International Conference on Web Engineering, p. 14. ACM (2006)
11. Noy, N.F., Musen, M.A.: Promptdiff: a fixed-point algorithm for comparing ontology versions. In: AAAI/IAAI, 2002, pp. 744–750 (2002)
12. Klein, M., Noy, N.F.: A component-based framework for ontology evolution. In: Workshop on Ontologies and Distributed Systems at IJCAI, vol. 3 (2003)
13. Ahmed, B., Driouche, R., Boufaida, Z.: Détection et représentation des relations entre versions d'ontologie (2009)
14. Chen, C., Matthews, M.M.: A new approach to managing the evolution of owl ontologies. In: SWWS, pp. 57–63 (2008)
15. Asif, H.K., Ahsan, S.M.: A critical analysis of the existing ontology evolution approaches. J. Am. Sci. **7**(7), 584–588 (2011)
16. Liu, K., Tang, S., Zhang, L., Tian, H.: A method based on RDB for detecting changes between multi-version ontologies. J. Comput. Inform. Syst. **8**, 3293–3300 (2012)

17. Franconi, E., Grandi, F., Mandreoli, F.: A semantic approach for schema evolution and versioning in object-oriented databases. Comput. Logic-CL **2000**, 1048–1062 (2000)
18. Noy, N., Klein, M.: Ontology evolution: not the same as schema evolution. Knowl. Inf. Syst. **6**, 428–440 (2004)
19. Djedidi, R., Aufaure, M.: Patrons de gestion de changements OWL. Ingénierie Des Connaissances, pp. 145–156 (2009)
20. Djedidi, R., Aufaure, M.: ONTO-EVOAL an ontology evolution approach guided by pattern modeling and quality evaluation. In: International Symposium on Foundations of Information And Knowledge Systems, pp. 286–305 (2010)
21. Keberle, N., Litvinenko, Y., Gordeyev, Y., Ermolayev, V.: Ontology evolution analysis with OWL-MeT. In: Proceedings of the International Workshop on Ontology Dynamics (IWOD-07), pp. 1–12 (2007)
22. Mahfoudh, M., Thiry, L., Forestier, G., Hassenforder, M.: Adaptation consistante d'ontologie à l'aide des grammaires de graphes. IC-24èmes Journées Francophones D'Ingénierie Des Connaissances (2013)
23. Dinh, T.: Grammaires de graphes et langages formels. (Université Paris-Est, 2011) (2011)
24. Liu, L., Ji, X., Li, L., Lü, S., Zhang, R.: A method for ontology change management based on belief revision. Int. J. Digit. Content Technol. Appl. **7**, 812 (2013)
25. Shafer, G., et al.: A Mathematical Theory of Evidence. Princeton University Press, Princeton (1976)
26. Tichy, W.: Configuration Management. John Wiley & Sons, Inc., Hoboken (1995)
27. Rajlich, V.: Modeling software evolution by evolving interoperation graphs. Ann. Softw. Eng. **9**, 235–248 (2000)
28. Rajlich, V.: A model for change propagation based on graph rewriting. In: Software Maintenance, 1997. Proceedings., International Conference On, pp. 84–91 (1997)
29. Sall, O., Thiam, M., Bousso, M., Lo, M.: Using Hoare's axiomatic semantics for checking satisfiability of ontology change operations. In: Information Science and Digital Content Technology (ICIDT), 2012 8th International Conference On, vol. 1, pp. 61–66 (2012)
30. Klein, M., Fensel, D., Kiryakov, A., Ognyanov, D.: Ontology versioning and change detection on the web. In: Gómez-Pérez, A., Benjamins, V.R. (eds.) Knowledge Engineering and Knowledge Management: Ontologies and the Semantic Web. EKAW 2002. LNCS, vol. 2473, pp. 197–212. Springer, Berlin, Heidelberg (2002). https://doi.org/10.1007/3-540-45810-7_20
31. Pepijn, R., Dean, M., Bench-capon, T., Shave, M.: An analysis of ontological mismatches: heterogeneity versus interoperability. In: Proceedings of the AAAI Spring Symposium on Ontological Engineering, Stanford, USA (1997)
32. Klein, M.: Change management for distributed ontologies (2004)
33. Sall, O., Thiam, M., Bousso, M., Lo, M., Basson, H.: A model for ripple effects analysis of cascading problems in ontology evolution. Int. J. Metadata Semant. Ontol. **5**(7), 171–184 (2012)
34. Sassi, N., Jaziri, W.: Types de changements et leurs effets sur l'évolution de l'ontologie. Actes Journées Francophones Sur Les Ontologies. (2007)
35. Sall, O.: Contribution à la modélisation de données multi-sources de type DATAWEB basé sur XML. (Université du Littoral Côte d'Opale, Université Gaston Berger de Saint-Louis) (2010)
36. Ashburner, M., et al.: Gene ontology: tool for the unification of biology.Nat. Genet. **25**, 25–29 (2000)
37. Grahne, G., Räihä, K.: Database decomposition into fourth normal form. In: Proceedings of the 9th International Conference on Very Large DataBases (1993)
38. Hamdi, F., Safar, B., Zargayouna, H., Reynaud, C.: Partitionnement d'ontologies pour le passage à l'échelle des techniques d'alignement. 9eme Journées Francophones Extraction Et Gestion Des Connaissances (2009)

39. Hu, W., Zhao, Y., Qu, Y.: Partition-based block matching of large class hierarchies. In: Mizoguchi, R., Shi, Z., Giunchiglia, F. (eds.) The Semantic Web – ASWC 2006. ASWC 2006. LNCS, vol. 4185, pp. 72–83. Springer, Berlin, Heidelberg (2006). https://doi.org/10.1007/11836025_8
40. Le Pham, T.A., Le-Thanh, N., Sander, P.: Some approaches of ontology decomposition in description logics. In: Loureiro, G., Curran, R. (eds.) Complex Systems Concurrent Engineering, pp. 537–546. Springer, London (2007). https://doi.org/10.1007/978-1-84628-976-7_60
41. Stuckenschmidt, H., Klein, M.: Structure-based partitioning of large concept hierarchies. In: McIlraith, S.A., Plexousakis, D., van Harmelen, F. (eds.) ISWC 2004. LNCS, vol. 3298, pp. 289–303. Springer, Heidelberg (2004). https://doi.org/10.1007/978-3-540-30475-3_21
42. Batagelj, V.: Analysis of large networks islands. Algorithmic Aspects Of Large And Complex Networks (1993)
43. Stuckenschmidt, H., Klein, M.: Reasoning and change management in modular ontologies. Data Knowl. Eng. **63**, 200–223 (2007)

The Effect of Digital Marketing on Micro, Small and Medium Enterprise in Indonesia

Alvin Igo Sasongko[1](✉), Gregorius Christian Widjaja[1], Jason Theodore[1], Nunik Afriliana[2,3], Tokuro Matsuo[4], and Ford Lumban Gaol[3]

[1] School of Information System, Bina Nusantara University, Jakarta, Indonesia
{alvin.sasongko,gregorius.widjaja,jason.theodore001}@binus.ac.id
[2] Faculty of Engineering and Informatics, Universitas Multimedia Nusantara, Tangerang, Indonesia
nunik@umn.ac.id
[3] Computer Science Department, Bina Nusantara University, Jakarta, Indonesia
fgaol@binus.edu
[4] Graduate School of Industrial Technology, Advanced Institute of Industrial Technology, Tokyo, Japan
matsuo@aiit.ac.jp

Abstract. Digital Marketing is a product marketing strategy using digital media and internet networks. Nowadays, digital marketing is a necessity for companies that thrive in the digital world. It greatly affects the growth of MSME (Micro, Small, Medium Enterprises) especially when it comes to increasing their sales performance by enhancing their brand awareness through various media or instruments such as a Facebook ad, Instagram advertising, Search Engine Optimization, and Search Engine Marketing. This research is aimed to measure how much impact of digital marketing utilization into Micro, Small, Medium enterprises. On the other hand, it also provides an understanding on what are the best digital marketing method to be utilized in the current situation. The method we used in this paper is quantitative. Data were collected through various questions constructed in an electronic questionnaire. This research provides a reference of digital marketing strategies for Micro, Small and Medium business owners. The result shows that digital marketing has a huge impact on the growth of Micro, Small and Medium Enterprise from increasing sales, increasing brand awareness, and gaining customers based on the data collected from both customer and Micro, Small, and Medium Enterprise players in Indonesia.

Keywords: Micro · Small and Medium Enterprise · MSME · Digital Marketing · Marketing Strategy

1 Introduction

MSME (Micro, Small, Medium Enterprises) has become one of the leading factors of economic growth in Indonesia. It has a huge contribution such as providing 120 million Indonesian with jobs and contributing by giving the country more than 60% of their

T. Matsuo et al. (Eds.): AIMD 2019, LNNS 677, pp. 147–156, 2023.
https://doi.org/10.1007/978-3-031-30769-0_14

GDP [1, 2]. As time goes by, more and more businesses, including MSME in Indonesia have shifted into the digital environment. It is proven by the emerging online business over the last few years. Creating a business account on Instagram, Facebook or any other social medias has become a standard in today's business environment, in fact, 70% of businesses has gone online by owning their own social media page across the globe [3].

The online competition getting tighter, business owners over the years have been looking for competitive advantages through social media features. Digital marketing is one of the most popular social media utilization. It is defined as the use of internet/digital media for marketing activities. Many MSME players in Indonesia have utilized this feature to enhance their business performance. Digital marketing comes with a huge variety such as Facebook ads, Instagram ads, Search Engine Optimization (SEO), Search Engine Marketing (SEM), and any other tools with their uniqueness and capabilities [4, 5].

The ability to adapt to the new way of marketing has become a very dominant factor and become one of the biggest sought in management skills. This is also particularly happened in the marketing field [6]. Over the years digital marketing has gained popularity across the world along with the increased popularity of social media and instant messaging such as Instagram, Facebook, and WhatsApp. A lot of people nowadays are connected to that platform. Many of them take advantage out of this situation since they realize how many people they can attract with digital marketing [7].

Digital marketing itself has some advantages as the data can be tracked easily which might be useful as insights to re-marketing the product and generating leads. This new approach of marketing comes at a relatively low cost compared to some conventional ways. Nevertheless, digital marketing also has cons since the technology and business trends are dynamic. Hence, the effectiveness of each digital marketing instrument needs to be re-evaluated [8]. Digital marketing still has a couple of disadvantages such as copyright problems, the customer's trust problem since customers can't really see the goods in real-time, and also trust in terms of the amount of fraud in the digital marketing world [9].

This research was aimed to give an overview on how much digital marketing can affect online SMSEs goals such as increasing brand awareness, number of sales, and customer engagement. Some of the most famous instruments utilized for digital marketing in Indonesia to be considered in this study are Instagram ads, Facebook ads, Tiktok and any other platform. It is conducted to understand the effectiveness of nowadays digital marketing platforms to MSME in Indonesia, by collecting data from various sources.

The research problem is the shifting paradigm of trend in marketing where nowadays business owners especially MSME players tend to go with digital marketing instead of traditional marketing. Through the research, we want to understand how much impact famous of digital marketing such as paid advertisement, SEO, and others alike can affect the MSMEs performance. Some important objectives in this study are:

- Evaluate the performance of digital marketing platforms when it comes to sales through analysis of consumer behavior.
- Evaluate the performance of digital marketing platform in terms of sales, engagement, brand awareness through business owner point of view.

- Evaluate the effect of digital marketing on brand awareness of MSME through consumer point of view
- Evaluate the most common digital marketing platform utilized by Indonesian MSMEs.

The main questions of the research conducted are stated below:

1. How much is digital marketing influence on MSME in Indonesia?
2. What is the most popular digital marketing platform to promote a business's products and services of MSMEs?

Digital Marketing is claimed to be a very useful method to increase sales, but there is also an opinion that digital marketing isn't important because it has a few drawbacks such as not being able to see directly who will be the seller and customer (many people used fake accounts) and only effective for the audience that often uses the internet. The relevance and importance of our research are to find out the truth about the effectiveness of digital marketing platforms that are available nowadays and also figure out which are the best methods to be used according to the need of MSMEs. Therefore, the hypothesis for our study are as follows:

H0: Digital Marketing has positive effects such as increasing sales, engagement, or brand awareness in Indonesia's MSME.
H1: Digital Marketing has no positive effects in increasing sales, engagement, or brand awareness in Indonesia's MSME.

We hypothesized that digital marketing has positive impacts on the growth of MSME by increasing their sales, brand awareness which is represented by the H0. The alternative hypothesis is that digital marketing has no positive effect on MSME in Indonesia, represented by H1.

2 Literature Review

MSME is defined by [10] as business organizations that contribute a significant economical role in a country. However, on the other side, they have limitations in term of human resources and financial support. As a result, they tend to resist the technological changes in their organization. MSMEs are encouraged to adopt digital technology to gain many advantages such as efficiency, profitability, and operational excellence for employees [11].

Digital Marketing is one of the new ways of people doing marketing. Its popularity emerged along with the increasing popularity of social media. Social media technology is getting more sophisticated every day. The differences between digital marketing and traditional marketing are in the kind of channel utilization. Digital marketing is performed mostly in digital channels such as online advertising, email marketing, social media, text messaging, affiliate marketing, search engine optimization, and pay-per-click. On the other hand, traditional marketing is performed in traditional channels such as television or newspaper [8].

Unlike in the past, nowadays most businesses are using digital marketing as their primary way of promoting their product despite the varied goals they have. Digital marketing is one of the most effective ways for people to increase their brand awareness, convert sales, and reach more engagement with the most effective and efficient ways possible. Digital marketing itself has some advantages as follows [12]:

- Support better communications with customers.
- Mobile access, meaning most people in the world have already connected through the internet and technology,
- Digital marketing allows business owners to reach their target market more easily.
- Affordability, digital marketing is also a cheaper alternative than traditional marketing.
- Easy tracking, meaning businesses have absolute control over data like the amount budget spent, how many people reached, and how many converted into sales.

The impact of digital marketing itself has been proven across the globe, including the Nike company in South Africa Nike has to shift their marketing approach to digital because of the customer behavior to the new way of marketing itself [13].

At the end of 2019, Indonesia ranked as one of the top 10 internet and social media most active users in the world. Therefore, digital marketing in Indonesia has become one of the emerging technological trends in recent years with many companies and enterprises shifting into digital. Many believe that digital marketing brings more benefits to their company by creating awareness to the public about the products, bringing traffic into the business page then converting it into sales [14].

Digital marketing has been going upwards over the last few years and it's predicted to go up even further with the rise of popularity of new platforms such as Tik-Tok. The competition of creating engaging campaigns for consumers is getting tighter. Digital marketers in Indonesia saw this opportunity and utilized their skills to take benefit by doing marketing agency up until the product supply chain [15, 16]. The public and government took notice of the ever-increasing popularity of digital marketing in Indonesia and saw an opportunity in this event. In 2020 Indonesia's economy has been hard hit by the COVID-19 pandemic. This situation leads to the condition of MSMEs players in Indonesia shifting their business to digital because of the social distancing. This made many businesses go digital and therefore it indirectly affected the digital marketing popularity even further [17].

3 Material and Method

This study adopted a quantitative method by conducting a survey. This quantitative research is proceeding by collecting and analyzing numerical data that have been gathered through the survey to find patterns, make predictions, find a causal relationship, and generalize the results of a wider population. The data were mainly collected from questionnaires.

The collected data are cross-sectional, within the one-week timeframe, by filling the google form. It was distributed between 2nd to 9th December 2020. The collected data are mainly from people who live in Indonesia in the age range from 18–30 years old.

The sample of populations is divided into two categories which are customers and the MSMEs players who utilize digital marketing. Sixty respondents from the customer side and 22 respondents from the business owner side have participated in this survey.

4 Result and Discussion

The data ware gathered from 60 participants of MSME in Indonesia who filled the distributed questionnaire. It was performed within a week. The research below shows graphs that represent consumer behavior in making online purchases because of advertising efforts from enterprises (Fig. 1).

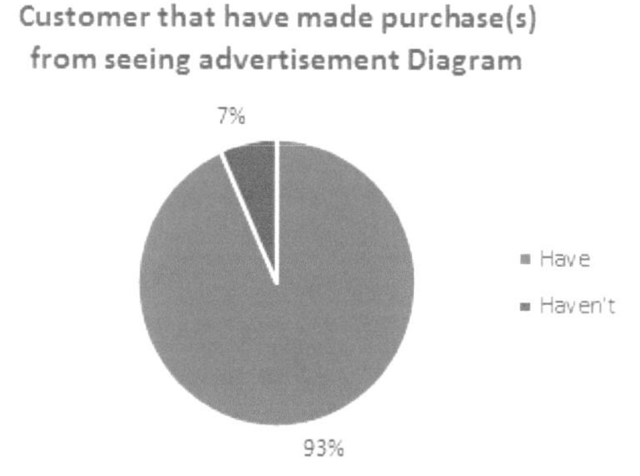

Fig. 1. Number of customers made purchases(s) by seeing an advertisement

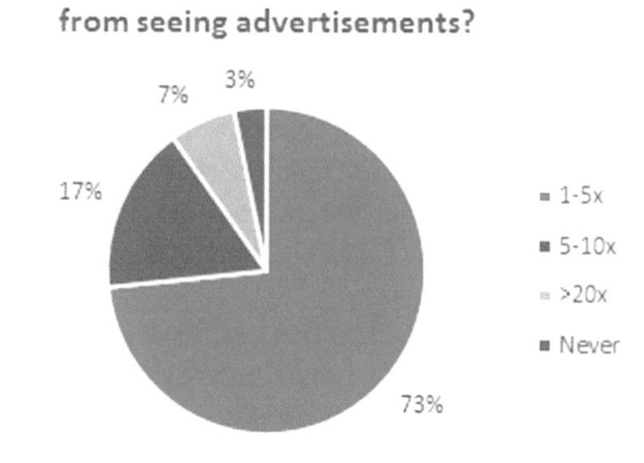

Fig. 2. The frequency of purchasing by seeing advertisement

The second pie chart as shown in Fig. 2 describes the number of times people take a part in online sales because of the marketing campaign created by enterprises. The result shows that people make online purchases mostly 1–5 times because of the marketing campaign by the amount of 73.3%. The second position is 5–10 times or 17%, followed by 7% who made an online purchase more than 20 times, and 3% who never made any online purchase. It is seen that 97% of the respondents have made an online purchase as they saw the digital advertisement.

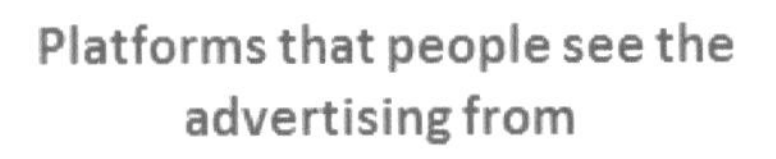

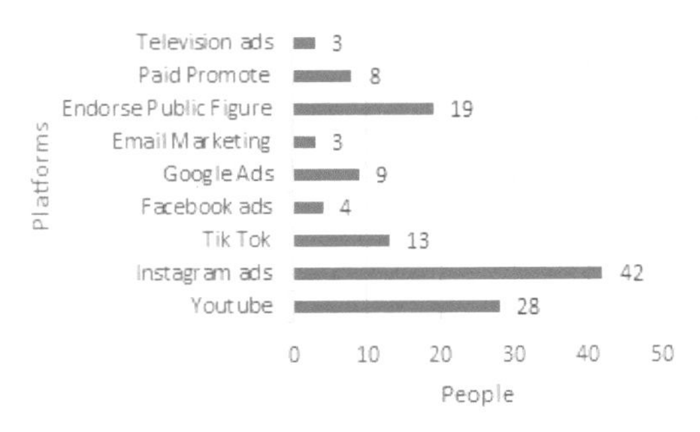

Fig. 3. Utilized platform to see advertisement

Figure 3 describes the media/platforms utilized to see the online advertisement and make their online purchase. Instagram dominated the utilization by taking a portion of 72.4% of the total people who make the online purchases, followed by YouTube advertising by 48.3%. The next highest sales conversion percentage are Tiktok advertising, it was around 22.4%, public figure endorsement 34.8%, paid to promote 13.8%, google ads 15.5%, Facebook ads 6.9% and then finally email marketing 5.2%. This graph represents the most effective digital marketing platform that converts best into sales.

Another important variable in digital marketing is brand awareness. Brand awareness is the measurement/degree of how familiar it is the enterprise's target audience to its product or simply the degree of how much people recognize the enterprise's product [10]. Here is some data to answer the question of doing digital marketing translates into good brand awareness and, if it is so, what is the best platform to increase brand awareness.

Figure 4 shows that 92% of total participants have come across a whole new product or are being introduced to brand new products because of the online advertising (Fig. 5).

Similar to the sales conversion, brand awareness effect from digital marketing mostly comes from Instagram advertising at 69.6%, followed by YouTube ads at 62.5% and the third one Google ads at 26.8%.

Shows how many people that aware of one's brand from ads

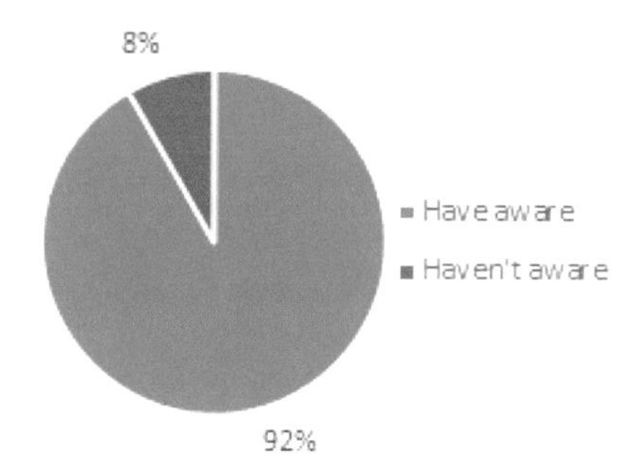

Fig. 4. Brand awareness as a result of online advertisement

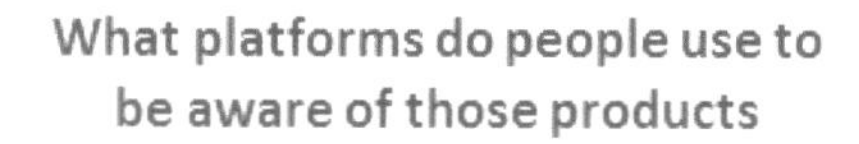

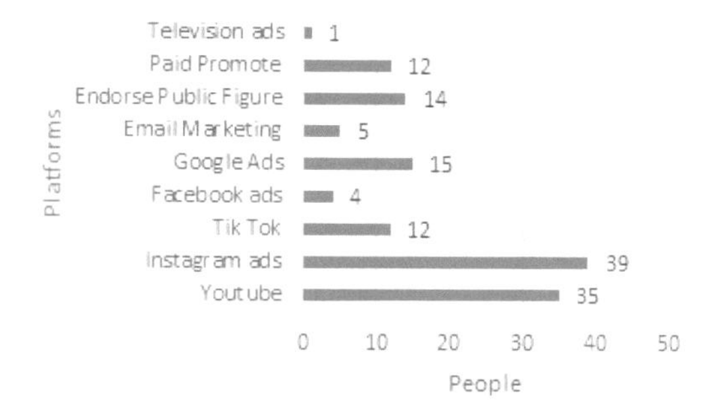

Fig. 5. Utilized digital marketing platform

In this section, we changed the respondents to MSME players who are also utilizing digital marketing platforms. The aim is to understand MSME player goals in utilizing the digital using digital marketing platform.

Derive from Fig. 6 and Fig. 7, most MSME owners utilized digital marketing to create brand awareness. It took place 86.4% of the audience. It is then followed by converting into sales at 81.8, and finally increasing engagement at 31.8% respectively. The most common platform used is Instagram ads at 77.3%, followed by public figures endorsement and paid promotes at the number of 27.3%.

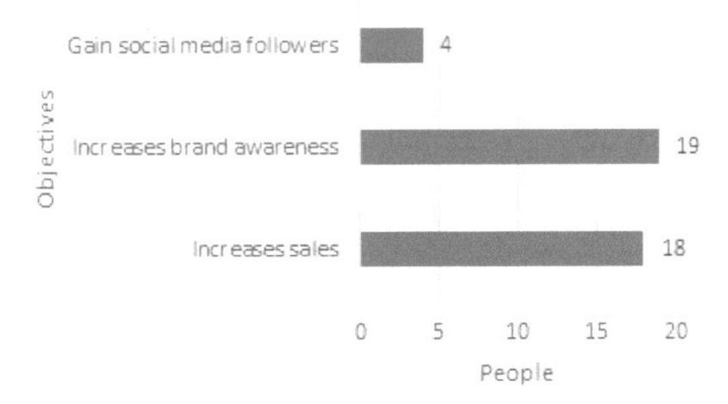

Fig. 6. The objective of digital marketing utilization

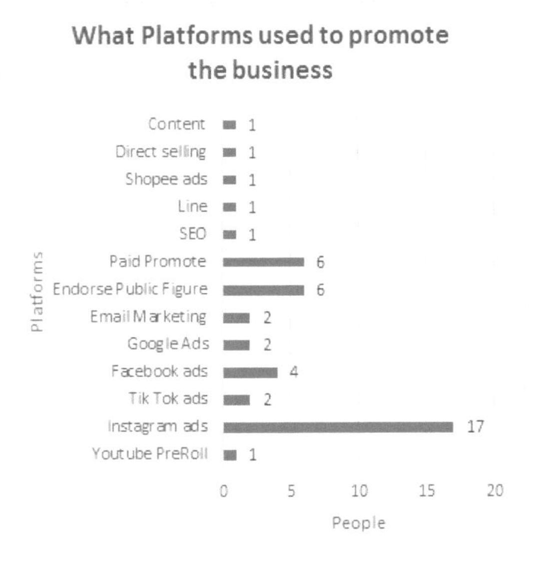

Fig. 7. Digital marketing platform utilized to promote the business

The last and might be considered to be the most important finding is the performance of the digital marketing platform. As shown on Fig. 8, most MSME owners (72.7%) agreed that digital marketing does help them increase their brand awareness, convert sales, and increase followers with a score of 4 out of 5. Some other respondents (18.8%), concluded that digital marketing is a perfect method to help them reach their goals in scaling up their business. However, around 9.1% of the respondent gave a moderate score in valuing the effect of digital marketing on their business.

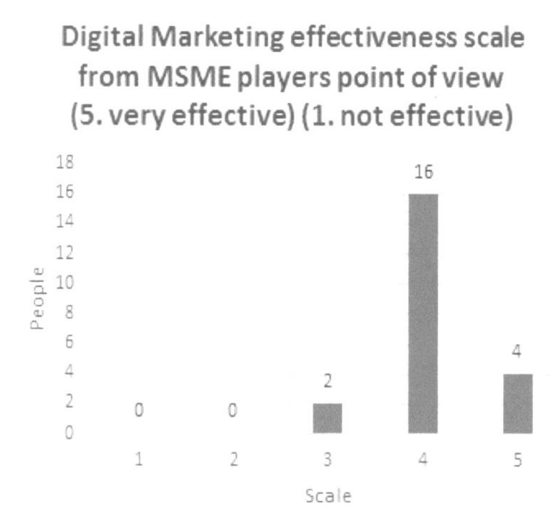

Fig. 8. Digital Marketing effectiveness scale from MSME players point of view

5 Conclusion and Recommendation

The study has collected the data and evidence from 60 consumers and 22 MSME owners from some different places across Indonesia. We can conclude and validate several things such as the fact that based on our findings, digital marketing does influence the MSME in a good way by allowing it to increase its brand awareness, convert into sales, and convert people into followers or loyal customers.

This finding validated the H0 hypothesis that stated Digital Marketing has positive effects such as increasing sales, engagement, or brand awareness in Indonesia's MSME. This is proven by the consumer behavior towards MSME advertising, with more than 90% of them either being introduced to a new product or making a purchase because of digital marketing. From the MSME owner's point of view, the finding stated that most of them are very satisfied with digital marketing with a score of 4 out of 5, as 5 is being very satisfied.

The data we gathered also confirms that so far Instagram Ads is the most effective digital marketing method. Instagram has proven to be very useful to be utilized for driving customers' intention of purchasing a product and creating brand awareness. More than 70% of respondents have been agreed with this conclusion. It is outnumbering other digital platforms.

This research will be beneficial for every enterprise's owner, by emphasizing the fact that digital marketing affects the marketing of enterprises in a good way. Therefore, every business owner needs to consider the method especially the one who wants to have a competitive advantage over their competitor by introducing their product to a larger customers.

References

1. Tambunan, T.: Recent evidence of the development of micro, small and medium enterprises in Indonesia. J. Glob. Entrep. Res. **9**(1), 1–15 (2019). https://doi.org/10.1186/s40497-018-0140-4
2. Micro, Small and Medium-sized Enterprises (MSMEs), no. September. Wellington Capital Adcisory (2021)
3. Yadav, M.: Social media as a marketing tool: opportunities and challenges. Indian J. Mark. **47**(3), 16 (2017). https://doi.org/10.17010/ijom/2017/v47/i3/111420
4. Taiminen, H.M., Karjaluoto, H.: The usage of digital marketing channels in SMEs. J. Small Bus. Enterp. Dev. (2015)
5. Warokka, A., Sjahruddin, H., Sriyanto, S., Noerhartati, E., Saddhono, K.: Digital marketing support and business development using online marketing tools: An experimental analysis. Int. J. Psychosoc. Rehabil. **24**(1), 1181–1188 (2020)
6. Valos, M.J., Ewing, M.T., Powell, I.H.: Practitioner prognostications on the future of online marketing. J. Mark. Manag. **26**(3–4), 361–376 (2010). https://doi.org/10.1080/02672571003594762
7. Bala, M., Verma, D.: A critical review of digital marketing. Int. J. Manag. IT Eng. **8**(10), 321–339 (2018)
8. Yasmin, A., Tasneem, S., Fatema, K., et al.: Effectiveness of digital marketing in the challenging age: an empirical study. Int. J. Manag. Sci. Bus. Adm. **1**(5), 69–80 (2015)
9. Todor, R.D.: Blending traditional and digital marketing. Bull. Transilv. Univ. Brasov. Econ. Sci. Ser. V **9**(1), 51 (2016)
10. Iswari, N.M.S., Budiardjo, E.K., Hasibuan, Z.A.: Integrated E-business system architecture for small and medium enterprises. In: Proceedings of the 2nd International Conference on Software Engineering and Information Management, pp. 240–243 (2019). https://doi.org/10.1145/3305160.3305193
11. Satvika Iswari, N.M., Budiardjo, E.K., Santoso, H.B., Hasibuan, Z.A.: E-business application recommendation for SMEs based on organization profile using random forest classification. In: 2019 International Seminar on Research of Information Technology and Intelligent Systems (ISRITI), pp. 522–527 (2019). https://doi.org/10.1109/ISRITI48646.2019.9034632
12. Chaffey, D., Ellis-Chadwick, F.: Digital Marketing. Pearson, UK (2019)
13. Reddy, G., et al.: Digital marketing impact on the consumer decision making process in Nike's customer retail operations in South Africa. University of Pretoria (2017)
14. El Junusi, R.: Digital marketing during the pandemic period; A study of Islamic perspective. J. Digit. Mark. Halal Ind. **2**(1), 15–28 (2020)
15. Adam, M., Ibrahim, M., Ikramuddin, I., Syahputra, H.: The role of digital marketing platforms on supply chain management for customer satisfaction and loyalty in small and medium enterprises (SMEs) at Indonesia. Int. J. Supply Chain Manag. **9**(3), 1210–1220 (2020)
16. Shaikh, F.A., Ali, M.: The effectiveness of social media marketing on consumer buying behavior: study of small medium enterprises (2017)
17. Rifai, Z., Meiliana, D.: Pendampingan dan penerapan strategi digital marketing Bagi Umkm Terdampak pandemi Covid-19. BERNAS J. Pengabdi. Kpd. Masy. **1**(4), 604–609 (2020)

Bidirectional Associative Memory as Normalisator Backpropagation Neural Network in the Signature Image Training

Fransisca Fortunata Dewi[1](✉), Ford Lumban Gaol[1], and Tokuro Matsuo[2]

[1] School of Computer Science, Binus University, Jakarta, Indonesia
fgaol@binus.edu

[2] Advanced Institute of Industrial Technology, Tokyo, Japan

Abstract. One of the technologies needed by humans, is supporting security technology in digital transactions that are present in the form of digital image signature verifier technology. Development of algorithms for the needs of signature verification is required to improve the performance (accuracy and speed) from the existing algorithms. This time, combination of Bidirectional Associative Memory (BAM) and BackPropagation Neural Network is used in the training process the signature in the form of digital images. Digital image signature through the purification process the data before processed with BAM and BackPropagation. To determine the effect of normalization BAM, carried out performance comparisons between the performance with BAM and the performance without BAM. From this study, it is evident that the use of normalisator can increase the speed of the training process digital image signature.

Keywords: Bidirectional Associative Memory · Backpropagation · Neural Network · Signature

1 Introduction

Along with the development of today's technology, the use of images as a verifier can be easily found. It is because of the use of images as a verifier is more secure than the use a passcode.

The use of images as a verifier can be found in many places, for example is the use of fingerprints and retinal imaging results on the electrical identification card. The use of images can also be found on scanning a passport photo to the database photo of fugitives.

Before the use of images as a verifier on the identity cards, the signatures that verified manually was enough to show the identity of the cardholder. Until today, the signature is still the one of component to verified. But, to improve services in the time of the service, we can develop various verifier signature algorithms. Besides of speed, the use of information technology in the signature verification process is expected to reduce human error.

T. Matsuo et al. (Eds.): AIMD 2019, LNNS 677, pp. 157–163, 2023.
https://doi.org/10.1007/978-3-031-30769-0_15

2 Related Research

According Srikanta et al. (2012), the signature verification process is a process to examine the signature to determine the validity of ownership. Before the use of technology in the signature verification process, still relied upon signature recognition process manually, which requires high accuracy and honesty. With the help of technology, the verification process become more effective and efficient, and can reduce the levels of human error.

Several algorithms have been developed for signature verification, including:

- Wavelet Packet (Romo, et al., 2009)
- HOG and LBP (Yilmaz, et al., 2011)
- Combination between Cross validation and Graph Matching (Bhole, et al., 2011)

Over time, many algorithms have been developed as an alternative to the signature verification process, such as a new algorithm, derived from the other algorithms, or the results of a combination of two or more algorithms. Therefore, the writer try to develop the two existing algorithms. The algorithm is selected for combined are Bidirectional Associative Memory and Backpropagation Neural Network. In this combination, Bidirectional Associative Memory serves as normalisator of signature image before trained by Backpropagation Neural Network.

According to Yash Pal Singh, et al (2009), Bidirectional Associative Memory (BAM) is a model of Neural Network are included in the category of Supervised Learning Networks as well as the Backpropagation, which is characterized by the presence of training mechanism. Bidirectional Associative Memory was first introduced by Bart Kosko based on previous studies and other models of associative memory.

According to Mohammed Waadalla, et al (2014), Neural Network is a collection of neurons that are connected via a line with their respective weights. Weights are kept updated in the process by a mathematical function. Meanwhile, according to Nader Salari, et al (2014), Backpropagation Neural Network is one of the most efficient methods to solve complex problems. That is because Backpropagation Neural Network can solve the complex problems to more simple structures and arranged adjacent side to not lose the description of complex systems previously. Backpropagation Neural Network is a neural network with multilayer feedforward concept consisting of an input layer, hidden layer and output layer. (Cao Jianfang, 2014).

3 Combination of Bidirectional Associative Memory and Backpropagation Neural Network

In this study, carried out the training process of a group of signature image, either without normalisator or with normalisator Bidirectional Associative Memory, by the method of Backpropagation Neural Network. Both of the training processes carried out in the same environment and, the same treatment. Committing the equal-treatment is:

- Value of learning rate coefficients = 0.5
- Binary Sigmoid activation function: $f(x) = \frac{1}{1+e^{-\sigma.x}}$

- The value of sigma (σ): cheating parameter = 1
- The number of nodes in the input layer = 9
- The number of nodes in the output layer = 1
- The number of nodes in the hidden layer = 7
- Maximum number of iterations = 20000
- Target error = 0.00001

For each image, subjected to treatment:

1. Purification of the image, which includes the process of binarization, noise removal, image scaling, image centralization, image rotation, and image trimming.
2. Pictures that have been purified, divided into 9 sections, forming a 3 × 3 matrix, such as Fig. 1. (Das & Dulger, 2009).

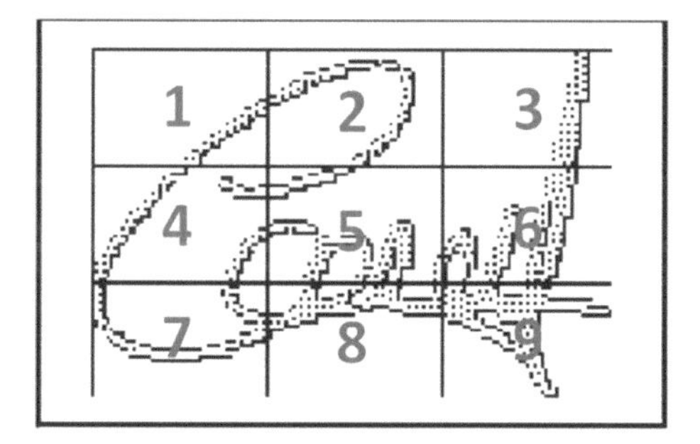

Fig. 1. Distribution of the signature image into 3 × 3 regions

3. At this stage, applyed different treatment between the side that use the normalisator, and the side that not using the normalisator.

   - Without normalisator: for each region, the calculated value of the ratio of black pixels to the total number of pixels in the region. Use the formula:

$$x = \frac{\sum black\ pixel}{\sum all\ pixel}$$

   with x = value comparison (Pansare, 2012)
   - With Bidirectional Associative Memory as normalisator:

     For each region, calculate value of the ratio of black pixels to the total number of pixels in the region. Use the formula:

$$x = \frac{\sum black\ pixel}{\sum all\ pixel}$$

   with x = value comparison (Pansare, 2012)

     Each region, divided into 4 equal parts, and is labeled a, b, c, and d, as shown in Fig. 2.

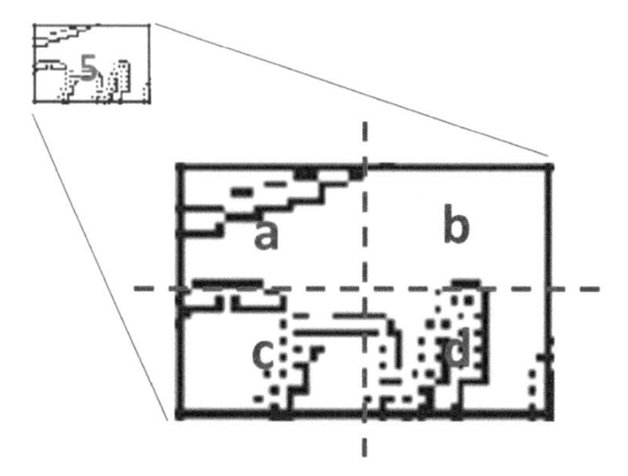

Fig. 2. Figure region which is divided into 4 equal parts.

With reference to the formation algorithms of matrix w for BAM storage, obtained matrix:

$$\begin{matrix} x_1 \\ x_2 \end{matrix}\begin{bmatrix} a & b \\ c & d \end{bmatrix}$$

$$y_1 \; y_2$$

$$w = \sum_{i=1}^{2} x_i^T y_i = \begin{pmatrix} a \\ b \end{pmatrix}\begin{pmatrix} a & c \end{pmatrix} + \begin{pmatrix} c \\ d \end{pmatrix}\begin{pmatrix} b & d \end{pmatrix}$$

$$= \begin{pmatrix} aa & ac \\ ab & bc \end{pmatrix} + \begin{pmatrix} bc & cd \\ bd & dd \end{pmatrix} = \begin{pmatrix} aa + bc & ac + cd \\ ab + bd & bc + dd \end{pmatrix}$$

Then calculated the determinant of each matrix, so we get a value for each region.

for matrix $A = \begin{pmatrix} a & b \\ c & d \end{pmatrix}$, so $\det(A) = ad - bc$.

The output of this stage for both of treatment is a 3×3 matrix that contains the identity of the signature image.

4. Determined a unique value as ID for the owner's signature. Training process is done for both treatment refers to a unique value as a target to be achieved.

4 Experiments

Time and number of iteration required for Backpropagation Neural Network to conduct the training process are use for analysis. The results of the analysis are summarized in Table 1.

From the results can be seen that the process of training with Backpropagation without normalisator BAM have a greater opportunity to experience the divergence compared with normalisator BAM.

Table 1. Variable measured for this study

	Average Time (second)	*Average Loop*
BAM and NN	0.01073725	578
NN	0.14653659	2138

With normalisator BAM, obtained percentage divergence:

$$\frac{4}{1041} * 100\% = 0.384245917387128\%$$

While no normalisator, obtained percentage divergence:

$$\frac{258}{1041} * 100\% = 24.7838616714697\%$$

Judging from the number of iteration needed backpropagation neural network to do the training, the result (Fig. 3):

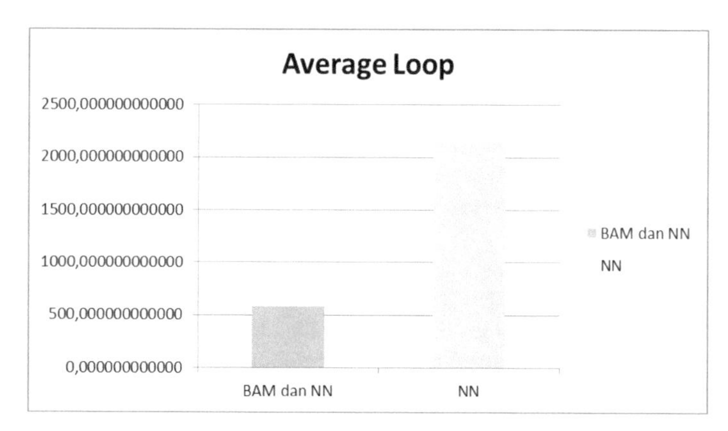

Fig. 3. Graph of comparison of the number of iteration is needed in the process of training

It is seen that the training process without normalisator require more iteration than the training process with normalisator. While the review of the time required Backpropagation Neural Network to conduct training, showed (Fig. 4):

Seen that without normalisator training process takes longer than the training process with normalisator.

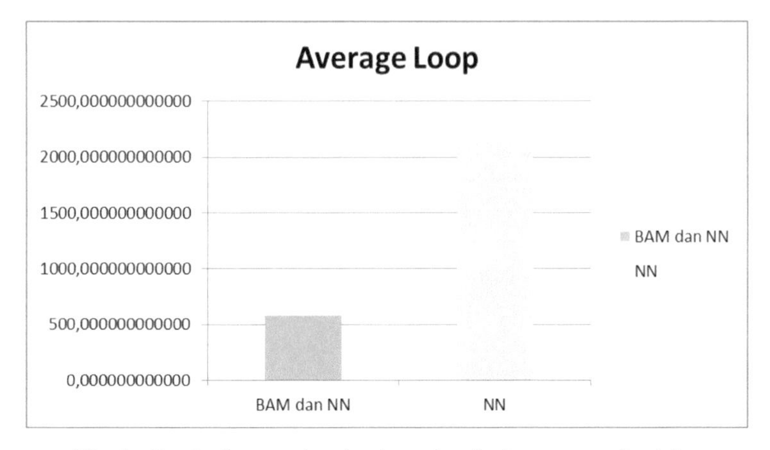

Fig. 4. Graph of comparing the time taken in the process of training

5 Conclusion

This research focuses on developing methods Bidirectional Associative Memory as normalisator training process of signature image with Backpropagation Neural Network method. The results showed, with the use of normalisator, the training of signatures images is faster 13 times, and also obtained a smaller divergence rate, at 0.384%, compared with no normalisator training, which reached 24 783%.

Although the results of the study has prove that the use of normalisator give the better results and faster process, there are some points that can be developed from this study, as follows:

- The use of the approach of the target value, other than those used in this study.
- The use of the signature images training method, other than Backpropagation.
- Increasing the efficiency of the complexity of the algorithms used in this study.
- The use of Bidirectional Associative Memory normalisator for other image recognition process, such as face and fingerprint.

References

Bhole, D.K., Vidhate, A.V., Velankar, S.: Offline signature using cross, validation and graph matching approach. Int. J. Technol. Eng. Syst. (IJTES) **2**, 67–70 (2011)

Cao, J., Chen, J., Li, H.: An Adaboost-backpropagation neural network for automated image sentiment classification. Hindawi Publ. Corp. – Sci. World J. **2014**, 1–9 (2014)

Das, M.T., Dulger, L.C.: Signature verification (SV) toolbox: application of PSO-NN. Eng. Appl. Artif. Intell. **22**, 688–694 (2009)

Martínez-Romo, J.C., Luna-Rosas, F.J., Mora-González, M.: On-Line signature verification based on genetic optimization and neural-network-driven fuzzy reasoning. In: Aguirre, A.H., Borja, R.M., Garciá, C.A.R. (eds.) MICAI 2009: Advances in Artificial Intelligence MICAI 2009. Lecture Notes in Computer Science, vol. 5845, pp. 246–257. Springer, Heidelberg (2009). https://doi.org/10.1007/978-3-642-05258-3_22

Pansare, A., Bhatia, S.: Handwritten Signature Verification using Neural Network. Int. J. Appl. Inf. Syst. (IJAIS) **1**(2), 44–49 (2012)
Salari, N., Shohaimi, S., Najafi, F., Nallappan, M., Karishnarajah, I.: A novel hybrid classification model of genetic algorithms, modified k-nearest neighbour and developed backpropagation neural network. PLoS One **9**, 1–50 (2014)
Singh, Y.P., Yadav, V.S., Gupta, A., Khare, A.: Bi directional associative memory neural network method in the character recognition. J. Theor. Appl. Inf. Technol. **2005–2009**, 382–386 (2009)
Srikanta, P., Blumenstein, M., Pal, U.: Automatic off-line signature verification system: a review. Int. J. Comput. Appl. **5**(2) (2012)
Waadalla, M., Yong, C.K., Yap, D.F., Rahim, R.A.: Underwater mobile robot global localization by using feedforward backpropagation neural network. Trends Appl. Sci. Res. **9**, 312–318 (2014)
Yilmaz, M.B., Yanikoglu, B., Tirkaz, C., Kholmatov, A.: Offline signature verification using classifier combination of HOG and LBP features. In: International Joint Conference on Biometrics, pp. 1–7 (2011)

Author Index

A
Afriliana, Nunik 37, 147
Azzahara, Devi Siti 37

B
Bakhoum, Ana 121

C
Christianto, Marven Immanuel 37
Cissé, Papa Alioune 121

D
Daisuke, Kasai 91
Dewi, Fransisca Fortunata 157
Diop, Ibrahima 121, 132

F
Fan, Ziran 1
Ferdiana, Ridi 111
Fujimoto, Takayuki 1, 22, 48, 59, 80

G
Gaol, Ford Lumban 14, 37, 147, 157
Gaye, Mouhamadou 121, 132

H
Hiromi, Miyauchi 91

K
Kuwahara, Nanami 48

L
Lostawika, Givbrela 37

M
Matsuo, Tokuro 14, 37, 147, 157

N
Nemoto, Taishi 59
Nugroho, Lukito Edi 111

R
Rahayu, Flourensia Sapty 111

S
Sasongko, Alvin Igo 147
Shimizu, Takashi 71

T
Tabei, Ken-ichi 101
Theodore, Jason 147

W
Watanabe, Yulana 22, 80
Widjaja, Gregorius Christian 147
Wijaya, Tommy 14

T. Matsuo et al. (Eds.): AIMD 2019, LNNS 677, p. 165, 2023.
https://doi.org/10.1007/978-3-031-30769-0

Printed in the United States
by Baker & Taylor Publisher Services